BIOMETEOROLOGICAL METHODS

ENVIRONMENTAL SCIENCES

An Interdisciplinary Monograph Series

EDITORS

DOUGLAS H. K. LEE
National Institute of
Environmental Health Sciences
Research Triangle Park
North Carolina

E. WENDELL HEWSON
Department of
Atmospheric Science
Oregon State University
Corvallis, Oregon

DANIEL OKUN
University of North Carolina
Department of Environmental
Sciences and Engineering
Chapel Hill, North Carolina

ARTHUR C. STERN, editor, AIR POLLUTION, Second Edition. In three volumes, 1968

L. FISHBEIN, W. G. FLAMM, and H. L. FALK, CHEMICAL MUTAGENS: Environmental Effects on Biological Systems, 1970

DOUGLAS H. K. LEE and DAVID MINARD, editors, PHYSIOLOGY, ENVIRONMENT, AND MAN, 1970

KARL D. KRYTER, THE EFFECTS OF NOISE ON MAN, 1970

R. E. MUNN, BIOMETEOROLOGICAL METHODS, 1970

BIOMETEOROLOGICAL METHODS

R. E. Munn

METEOROLOGICAL SERVICES OF CANADA
TORONTO, ONTARIO, CANADA

1970

ACADEMIC PRESS   New York and London

ACADEMIC PRESS, INC.
111 Fifth Avenue, New York, New York 10003

United Kingdom Edition published by
ACADEMIC PRESS, INC. (LONDON) LTD.
Berkeley Square House, London W1X 6BA

LIBRARY OF CONGRESS CATALOG CARD NUMBER: 71-97488

PRINTED IN THE UNITED STATES OF AMERICA

CONTENTS

9.   Physical Methods: Illustrative Examples

10.   Synoptic Applications

11.   Seasonal Relationships

12.   Studies of Past Climates

13.   Climatic Classification and Indices

14. Engineering and Economic Applications

Appendix. Problems

References

PREFACE

Biometeorology is the study of relationships between weather and life. Plants, animals, and people respond in complex ways to the atmospheric environment. Biometeorology inquires into the nature and effects of short-term stress and it speculates about long-term climatic adaptation.

This book is not a survey of biometeorology. Specialized reviews already exist. Because of the multidisciplinary nature of the subject, however, there is a need for a unified look at methodology. Examples of biometeorological studies have been chosen, therefore, not because the results are necessarily significant but because the method is instructive. Hopefully, the book is controversial in a positive sense.

A brief survey of biometeorology is given in Chapter 1 to orient the reader approaching the subject for the first time. The remainder of the book seeks to place in perspective the various experimental, empirical, analytical, and physical methods that are being used or could be used in biometeorology.

The effect of weather on life is inevitably a very diverse subject of inquiry. Results are appearing in many unrelated scientific journals, and quite similar problems are approached in rather different ways, using different terminologies. I believe that a single author can produce a more integrated "story" than can 14 specialists, each writing a separate chapter.

No previous book on biometeorology has been written from this point of view. The present volume should be of value to anyone seeking assistance in the design of experiments and analysis of environmental data. The analogies found in other scientific disciplines may not be precise but the methods may be suggestive. The book contains examples drawn from the fields of meteorology, hydrology, physiology, ecology, biology, medicine, air pollution, geography, forestry, agronomy, and engineering. A layman's familiarity with meteorology and a knowledge of elementary statistics are the only prerequisites for most chapters.

ACKNOWLEDGMENTS

I am indebted to many people for assistance. Dr. Peter Barry took the time to rewrite my first draft of Section 2.3, and I have accepted his revision gratefully. Others who have assisted with individual sections include Dr. F. Fanaki, Dr. H. E. Turner, Dr. B. Prasad, J. A. W. McCulloch, V. S. Derco, Dr. W. Ball, Dr. J. Clements, H. Cameron, G. R. Kendall, Dr. W. Baier, W. C. Palmer, E. I. Mukammal, and M. K. Thomas. Last but not least I must thank Dr. H. E. Landsberg, who read the entire manuscript, and Mr. E. Truhlar, who provided a sentence-by-sentence editorial review.

I wish to dedicate this book to my wife Joyce, whose patience and understanding are matched by a keen sense of proportion and good humor.

BIOMETEOROLOGICAL METHODS

1/WEATHER AND LIFE

1.1. PHYSIOLOGY, ECOLOGY, AND BIOMETEOROLOGY

The atmosphere is a life-sustaining reservoir of oxygen for man and animals, of CO_2 for vegetation, and of water for all forms of life. Through evolution, living organisms have adjusted remarkably to this life-giving but sometimes harsh environment. Man is the prime example, but desert insects burrow into the sand to escape the heat of the day while arctic animals insulate themselves from extreme cold by tunneling into snow. The single plant, stunted because of wind, heat, drought, or cold, may flourish if other plants are established at the same location, changing the microclimate. A further example is the pine tree, whose cone protects the seed during the intense but short-lived heat of forest fires; after such a calamity has occurred, however, the heated cone swells and cracks, permitting the seed to scatter in the next few days, a first step towards forest regeneration.

An interesting speculation is whether living organisms are in a "delicate balance" with their environment or whether they can adjust to large swings of the climatic pendulum. At what stage in overgrazing of grassland does the area become a dust bowl and change to desert? What is the effect on fish of a change in water quality? At what concentration does atmospheric pollution become harmful to man? These are typical questions that one seeks to answer through the sciences of *physiology* and *ecology*.

Physiology is the study of the ways in which a living creature stays alive and grows. A physiologist examines the intricate relations between the various cells and tissues of the organism, the organization of the parts into the whole, and the responses to environmental stimuli. *Ecology*,

1

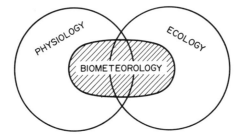

FIG. 1. Physiology ↔ Biometeorology ↔ Ecology.

on the other hand, is the study of relations between populations of living organisms and their environment, which include other living organisms.

A typical ecosystem is the forest clearing, in which competing species of grasses, shrubs, and tree seedlings interact with each other and with a number of weather factors in a complex way. These interactions determine whether forest regeneration will take place and if so, how quickly it will occur.

Biometeorology is that part of physiology and of ecology (Fig. 1) in which the atmosphere (natural or man-made) is a significant environmental factor; the living organism is then said to be weather sensitive. Determination of the area over which a beaver forages for trees for its dam is an interesting ecological question but not a biometeorological one. The release of ragweed pollen, on the other hand, depends on weather factors such as humidity and sunlight; in addition, the subsequent delivery of pollen from source (ragweed) to receptor (hay fever sufferer) is by wind transport; finally, receptor response is partly dependent on atmospheric factors such as temperature and humidity. Biometeorology is therefore important at three stages in this ecological chain.

1.2. AN EXAMPLE OF A BIOMETEOROLOGICAL STUDY

The general nature of a biometeorological investigation is illustrated in a study by Linzon (1). The Canadian Pacific Railway Company has operated a roundhouse at Chalk River, Ontario, Canada since 1912. Smoke from the steam locomotives has been blowing across a stand of jack pine for many years. The trees at the time of the investigation (1959) were all about 45 years old and were growing on a fairly uniform sandy soil. Visual inspection indicated that the trees near the roundhouse were not as large as those some distance away. Typical samples were cut

TABLE I

THE HEIGHT AND DIAMETER GROWTH OF JACK PINE TREES IN LOCATIONS
ADJACENT TO AND AT A DISTANCE FROM CHALK RIVER ROUNDHOUSE[a]

	Age at stump height (years)	Total height (m)	Diameter at breast height (cm)
Adjacent to roundhouse	44	8.0	12.2
	47	7.1	14.0
	43	8.5	12.4
Approximately 1.6 km	44	16.5	20.8
from roundhouse	44	16.4	21.3
	40	16.8	19.8

[a] Linzon (1).

and examined, some results of which are given in Table I, which merely confirm the visual observations that trees near the roundhouse were stunted, presumably because of smoke or SO_2.

Locomotive dispatch figures were available from the year 1927 onward. These are given in Table II, together with the annual radial growths obtained from tree-ring measurements. The locomotive figures reflect the economic state of Canada during the period of record. Following the boom in the 1920's, there was a slump in railways operations in the 1930's. During the war and postwar years of the 1940's, the average monthly dispatch of engines exceeded that of the 1920's.

In the trees some distance from the roundhouse, there was a decreasing rate of annual wood increment (4.5 mm in the 1920's to 1.7 mm in the 1940's); this is caused by another biometeorological factor—the intense competition within the maturing stand for light and soil moisture. The trees near the roundhouse, on the other hand, showed suppressed growth in the 1920's and 1940's but an intermediate period of recovery in the 1930's when railway operations were low. According to Linzon, such a growth pattern in jack pine is very rare. Typical tree cuttings are shown in Fig. 2.

These data strongly suggest but do not prove that there is a connection between growth rates and roundhouse activity. To demonstrate the existence of a cause-and-effect relation, detailed physiological experiments are required in order to determine the response of a jack pine to steam locomotive emissions. One might be interested in knowing, for example, whether it was the deposited soot or the SO_2 gas that was affecting the trees. The initial biometeorological investigation is, however, useful in indicating what further studies may be fruitful. In this particular

TABLE II
RELATIONSHIP BETWEEN RADIAL GROWTH OF JACK PINE TREES AND ENGINE
DISPATCH FROM ROUNDHOUSE AT CHALK RIVER, ONTARIO[a]

Year or period	Radial growth (mm)		Average monthly dispatch of engines[b]	Average monthly rainfall May–Aug. (cm)[c]
	Away from roundhouse	Near roundhouse		
1921				
1922	3.3	0.5	—	—
1922	3.0	1.1	—	—
1923	4.5	0.8	—	—
1924	5.3	0.8	—	—
1925	4.8	1.3	—	—
1926	6.0	1.5	—	—
1927	5.8	0.8	447	—
1928	5.0	0.5	464	—
1929	3.3	0.8	436	—
1930	4.3	0.8	402	—
1921–1930 (average)	4.5	0.9	437	—
1931	4.5	1.8	327	5.1
1932	4.5	2.8	301	7.9
1933	3.8	3.8	290	8.5
1934	3.3	3.8	217	6.2
1935	3.3	3.8	207	8.7
1936	3.3	3.5	231	9.5
1937	2.8	4.0	255	5.8
1938	3.0	3.8	241	6.9
1939	2.5	3.0	219	9.5
1940	1.5	2.8	298	10.7
1931–1940 (average)	3.2	3.3	258	7.0
1941	1.5	2.8	461	4.2
1942	1.8	2.3	534	6.9
1943	1.8	1.5	657	8.3
1944	1.8	0.8	638	6.5
1945	2.3	0.8	703	5.9
1946	2.0	0.8	694	7.4
1947	1.5	0.5	716	9.3
1948	1.5	0.2	620	7.7
1949	1.3	0.5	530	7.5
1950	1.3	0.8	476	8.7
1941–1950 (average)	1.7	1.1	603	7.2

[a] Linzon (1).

[b] Engine dispatch records available from 1927 onward. The lowest number of engines serviced was 179 in June, 1939, and the highest number was 848 in March, 1946.

[c] Rainfall records available from 1931 onward.

Fig. 2. Discs removed at butt level from jack pine trees in 1959, showing annual wood increment. The larger disc is from a normal tree in the Chalk River area while the smaller disc is from a tree near the locomotive roundhouse (1).

example, the presence of the roundhouse was perhaps obviously related to the stunted growth of the jack pines. In other cases, where a large number of factors may possibly be involved, the interactions may not be so evident.

1.3. REPRODUCIBLE AND SPECIFIC RELATIONS

The atmosphere is a fluid which obeys physical laws. This permits the use of mathematical models for prediction, provided that the assumptions in the models are satisfied in the real atmosphere. Some aspects of the biological sciences are capable of mathematical modeling but frequently the interactions of a living organism with its environment are so complex that only a statistical relation can be expected. For example, only some people are affected by asthma; even among those who are, the reaction to a given set of environmental conditions is variable. This is so even for a single individual whose response may depend upon whether he is tired or rested, tense or relaxed.

As another complication, living organisms sometimes develop a tolerance to stresses. Although there may be a readily identifiable reaction to a single temperature extreme or to a high concentration of pollution, repetition may diminish the effect. Thus, the same results are not produced by a given set of initial conditions, unless physiological factors are included.

The physical scientist demands that laboratory experiments be designed so that:

(a) The results are reproducible.

(b) The results are specific.

(c) The results are describable coherently in a physical model (this sometimes comes later).

These three conditions are not readily met in the atmosphere because of the difficulty in performing a "controlled" experiment. Vertical variations of wind, temperature, humidity, and pollution are often great, even within 10 m of the ground; horizontal variations are also large in and near cities, forests, valleys, and shore lines. Because a suburban commuter moves from one microclimate to another, as does a forest animal, or a bird, interactions are difficult to trace unless the response is associated with a very short but intense stress. Plants, being stationary, present fewer experimental problems; even so, the microclimate at the top of a forest canopy differs from that at the forest floor. When time variability and physiological factors are added (e.g., the sensitivity of a tobacco leaf to oxidant damage is dependent on its stage of development), the task facing a biometeorologist can be appreciated.

Finally, it should be noted that atmospheric stresses are not necessarily additive. A *synergistic* response is one in which the combined effect of two or more stresses is greater than the sum of the responses to each stress operating separately. In other cases, the response is said to be *antagonistic* or competitive, with a reduction in the total response. Buell and Dunn (2) suggest that it is not yet clear whether urban pollution and cigarette smoking are synergistic or competitive factors in the development of lung cancer; the question is of considerable importance in the design of further epidemiological studies.

1.4. READJUSTMENT OF LIVING ORGANISMS TO THEIR ENVIRONMENT

It is convenient to consider environmental stress of short duration (a few minutes to a few days) separately from long-term influences (a few months to years). In neither case is stress necessarily harmful. Mountain climbing by a lowland man changes his breathing rhythm and body heat balance temporarily; although the exercise may be tiring, it is also invigorating. Living for a longer period at low atmospheric pressures, as in the High Andes, leads to more permanent physiological adjustments.

A *meteorotropism* is a change in an organism (or a large population) that is correlated with a change in atmospheric conditions. The example given in Fig. 3 (3) shows the 1952 weekly death rates from respiratory and heart diseases in London Administrative County, England. Light winds and a temperature inversion resulted in high pollution levels during the period December 5–9 when there was also a rise in death rates. Not all meteorotropisms are this marked and some are controversial, particularly when there is no apparent physical explanation.

Another example of short-term stress is the temperature extreme. Mammals must maintain a nearly constant body temperature of 37°C. When heat loss becomes too great and there is a sensation of chill (see Chapter 9), three physiological reactions begin (4). First, the blood vessels to the skin contract, reducing the flow of warm blood to the surface. Then, shivering involuntarily increases heat production. Finally, body hair stands on end, providing a thicker insulating layer for those mammals that are furry. A longer-term temperature decrease, as from summer to winter, leads to a reduction in blood volume.

When there is an excess heat load, on the other hand, an increase in breathing rate (panting) and the evaporational cooling of sweat help to reduce body temperature. Leaves also regulate their daytime heat load to a certain extent by transpirational evaporative cooling. Some plants change their leaf orientation to present a minimum surface towards the direct sunlight in hot weather. Butterflies also use this thermoregulatory mechanism.

Another stress on vegetation is soil moisture depletion. The plant adjusts progressively by closing its stomata (reducing transpirational moisture loss) and extending its roots deeper into the soil. On sunny days in dry climates, some species undergo a regular decrease in evapotranspiration shortly after noon due to temporary moisture stress (5), with recovery later in the afternoon.

Environmental stresses acting over many years or centuries may lead to *acclimatization* with permanent changes in physiology or behavior. The theory of climatic determinism asserts that life forms are very much the product of environment, although more so historically than in the present age of the controlled climates of houses (humans), barns (farm animals), and greenhouses (plants).

Physiological adaptation to stress is common but differing pathways may be used to achieve the same goal. The aborigines of Australia are able to sleep comfortably on the cold ground, although they lose more body heat than ordinary people (4). The Indians of Tierra del Fuego also

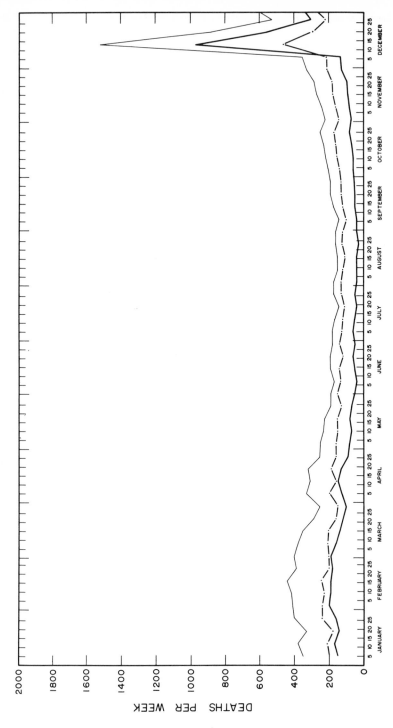

FIG. 3. Weekly deaths in London Administrative County, England during the year 1952. Severe smog occurred December 5-9 (3). (——) four respiratory diseases; (-•-) three heart diseases; (——) deaths due to both causes.

live with little covering but they have a high heat production rate and their body temperature falls only slightly in cold weather (4).

Desert insects tend to have long legs, which permit them to keep their bodies off the hot sand (6). All insects contain so little water they would soon dry out if evaporation were used for cooling; therefore, many of them have a protective wax coating.

There are many illustrations of acclimatization in polar regions. One example has been discussed by Dadykin (7); he found a substantial reduction in reflection of sunlight, particularly in the far-red and near-infrared part of the spectrum (as much as 50%), from leaves of plants grown in the arctic or in artificially cooled soils, as compared with plants in temperate latitudes.

The climatic warming of the 1920's and 1930's in northern Europe may have had an effect on bird ranges, as suggested by Harris (8). For example, starlings were first observed in Iceland in 1935, the fieldfare appeared in Greenland and Jan Mayen Island in 1937. Harris emphasizes that the connection between bird ranges and changing climate need not be direct or immediate. A complete ecosystem is involved; food must be available at the new location (insects and vegetation), while if conditions remain satisfactory in areas where birds have nested for generations, there may be a time lag of more than a decade before a northward movement begins.

A living organism is not always able to acclimatize to long-term stress; however, it may take many years before a harmful effect can be detected. The population of a fluorspar mining community in Newfoundland has been studied by de Villiers and Windish (9). Deaths from lung cancer during the period 1950–1961 among the miners of the town of St. Lawrence (population of about 2000) were 29 times greater than would be expected from the rates for all of Newfoundland. In addition, there was a shift to lower age groups at St. Lawrence, a finding that appeared to be related to the length of time since the miner had first worked underground, suggesting an occupational factor. Figure 4 shows the age at death versus the age at first underground exposure. The average induction time was 19.1 years. Subsequent environmental testing in the mines showed that air concentrations of radon and daughter products were 2.5–10 times the internationally recognized maximum permissible levels.

The methodology of the Newfoundland study is of interest. Many of the miners did not develop lung cancer but the incidence of those that did was much higher than for a comparable control group. A relation between the age at death and age at first underground exposure was established. Finally, an occupational hazard (high radon concentrations

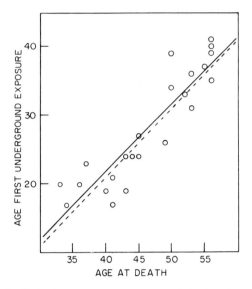

FIG. 4. Age at death by lung cancer versus age at first underground exposure of miners in St. Lawrence, Newfoundland. Solid line is the regression of y on x; broken line is the predicted age at death based on an average induction time of 19.1 years (9).

in the mines) was determined. The evidence is convincing, although not conclusive. This is typical of many environmental studies.

Climatic factors are often used to explain the patterns of life. Conversely, biological evidence sometimes makes it possible to infer the climate. Spore distributions in subsoils and fossils may be useful on the very longest time scales (paleoclimatology), tree rings assist in studies of climatic variations during the last few centuries, and vegetation patterns are often valuable indices of present-day climate, revealing the presence of very local frost pockets, wind circulations, or pollution sources.

1.5. READJUSTMENT OF THE ENVIRONMENT BY LIVING ORGANISMS

The microclimate is changed greatly by man, by vegetation, and, on occasion, by animals and insects. Cattle sometimes overgraze grassland, beavers build dams to flood low-lying ground, and locusts denude regions. Less dramatic but more significant is the effect of vegetation on temperature, humidity, and wind near the ground (10, pp. 155–166). After a world-wide warming trend over decades, the forest advances into the subarctic by modifying the microclimate near its forward edge. Temperature extremes in the undergrowth are not as severe as those over barren ground during the growing season, and there is less wind, permitting the tree line to creep north. In the arctic itself, as noted by Larsen (11, p. 10), vegetation often causes the subsoil to remain cool and damp, although air temperatures within the plant cover are higher than those over the tundra.

Ecologists are interested in determining whether world vegetation patterns are in equilibrium with their environments and whether existing equilibrium communities are most efficient in terms of productivity and water use. Ecologists are concerned, for example, about the effect of cutting the forest near the tree line in the subarctic or in mountain ranges. The climatic balance is so delicate that many hundreds of years may be required for the area to recover from the effects of the changed summer frost patterns and winter gales.

Forests have a significant effect on the soil moisture, the water table, and the timing and intensity of spring runoff from melting snow. Except very near the surface, the soil beneath a forest is drier than that below a grassy meadow in the growing season; this is due to the deep penetration of tree roots and the large leaf area for transpiration. An average maple tree has a total leaf area of about 4×10^3 m (1 acre). After a forest has been cut down, therefore, the area may turn into swamp because of the rising water table. As another example, Dounin-Barkovsky (12) refers to "non-beneficial vegetation and water resources." He believes that cane undergrowth along desert irrigation canals is a source of great moisture loss, as much as five times that from an open water surface.

Man, of course, greatly modifies his environment. His cities, for example, change the wind flow, the heat and moisture balances, and the air quality. He reshapes the world vegetation patterns, introduces shelter belts, irrigates the deserts, and turns valleys into lakes. He has created controlled climates in houses and factories. Even ragweed pollen, which is usually classified as a natural pollutant, only occurs in high concentrations as a result of man's activities—the ragweed flourishes in sterile subsoil that has been exposed in the construction of buildings, highways, and railway tracks.

Not all human intervention is detrimental, of course. In the Okanagan Valley of the Rocky Mountains, for example, where fruit trees are grown on terraces that extend up the valley sides, some of the orchard land is slightly concave, increasing the very local danger of frost. Man-made climatic amelioration has been achieved in this case by using heavy earth-moving equipment to cut air drainage channels to lower elevations.

1.6. THE MULTIDISCIPLINE APPROACH

The need for multidisciplinary cooperation in biometeorology has been emphasized many times, e.g., by the Study Group on Bioclimatology, of the American Meteorological Society (13). It is much easier, however, to recommend cooperation than to achieve it.

A university community often includes environmental scientists scattered through many departments—geography, physics, mathematics, medicine, botany, zoology, engineering, soil science, chemistry, occupational health, and even economics or anthropology. These scientists are frequently interested in overlapping parts of the same larger problem but sometimes are hardly aware of each other's existence. They may attend different scientific meetings, read different journals, and find themselves unable to communicate except at the layman's level.

This situation is changing slowly with the formation of broadly based Institutes of Environmental Studies and with the development of national and international interdisciplinary programs such as the IBP (International Biological Program) (13).

The brief outline of biometeorology given in this chapter indicates the wide scope of the subject. It might be argued that environmental studies are becoming less important in this age of controlled climates. However, new challenges arise because of population growth in arctic and tropical regions, the urgent problem of agricultural productivity, a realization of the importance of environmental factors on human efficiency and sense of well-being, the appearance of new stresses such as oxidant smog, and the need for intelligent planning of conservation and recreational areas.

2/SAMPLING OF THE ATMOSPHERE: TIME CONSIDERATIONS

2.1. TIME VARIABILITY IN THE ATMOSPHERE

In the depths of the ocean, the environment changes little from day to day. Throughout the lower atmosphere, on the other hand, the weather varies widely in both space and time; the heat balance at the ground is disturbed when a single cloud moves across the sun.

Astronomical factors impose daily and annual periodicities on averaged meteorological data. (Air temperature is usually higher at noon than at midnight, and in summer than in winter.) Even the most sharply tuned cycles, however, are rather unreliable predictors on any particular occasion. On some winter days in temperate latitudes, for example, the temperature may be higher before sunrise than at noon, if a cold front passes. As another example, the onset of the rainy season in Panama varies from year to year; the solar clock cannot be used to predict the arrival date exactly.

Biological data display daily and seasonal cycles. These are regulated by the meteorological cycles, but the relationships are not precise. Transpiration and CO_2 intake of plants depend partly on physiological factors, as does the release of spores and pollen.

Air pollution emissions in cities have not only daily and seasonal, but also weekly rhythms. These interact with the meteorological periodicities to produce air quality cycles and also to modify slightly the meteorological cycles. There is a tenuous connection, for example, between the reduced pollution emissions on weekends and rainfall. A classical study of precipitation in Rochdale, England by Ashworth (14) showed 13%

13

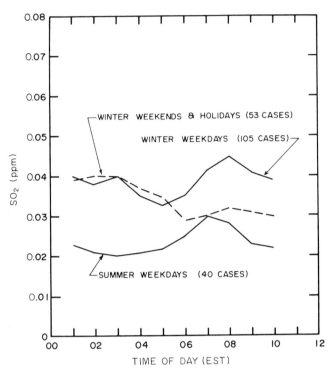

F IG. 5. Average SO₂ values at a sampling station in Ottawa, Canada for the hours from 0100 to 1000 EST on days when the surface wind during that period was ≳5 m/sec and there was no inversion. Insufficient summer weekend cases available. Ottawa shifts to daylight saving time in summer.

less rainfall on Sundays than on weekdays over a 10-year period, although there is the possibility of experimental bias; the observer may have occasionally neglected to empty the rain gauge on Sundays.

In areas which change from standard to daylight time in summer, there is a very precise shift of 1 hr in the diurnal cycle of urban activity and pollution emissions. An apparent effect on SO₂ concentrations at a sampling station in Ottawa, Canada is shown in Fig. 5. To eliminate the influence of meteorological processes such as the morning breakup of a nocturnal temperature inversion, the timing of which changes with the season, the data have been restricted to days when the surface wind was 5 m/sec or higher during the period 0100–1000 EST and there was no inversion. The weekly morning peak is an hour earlier in summer (0700 EST) than in winter (0800 EST). Furthermore, the peak is indistinct on weekends and holidays. Figure 5 suggests, but does not prove, a relation between human activities and pollution levels.

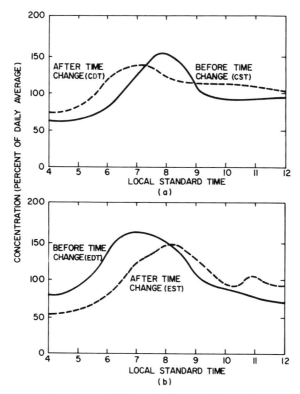

FIG. 6. Average hourly values of CO for the five weekdays prior to, and following, the change from standard to daylight time, years 1963–1965 inclusive (15). (a) Spring season change in Chicago; (b) autumn shift in Washington, D. C.

Figure 6 provides another example (15, p. 13). Average hourly values of CO are plotted for the five weekdays prior to and following the local time changes in Chicago and Washington. Note that the ordinates are given as a percentage of the daily average concentration of CO, an example of a normalization technique designed to counteract, at least partially, the day-to-day meteorological variability. There is a shift of 1 hr in the morning peak of CO, presumably associated with an equivalent shift in the time of the automobile traffic peak. In contrast, at Cincinnati, which did not change to daylight saving time during the years 1963–1965, there was no change in the time of occurrence of the peak.*

Daily, weekly, and annual cycles have direct physical explanations

*An alternative approach is to correlate CO concentrations with traffic densities. However, if there are many automobiles moving very slowly, traffic counters record small numbers whereas CO emissions are relatively high.

and are often well defined when environmental data are averaged. Other periodicities are more tenuous or irregular. For example, weather systems may cross a region at about 4-day intervals but after a few weeks, the rhythm may change. A central problem in biometeorology is the separation of the regular from the random parts of the time variability. In the language of the communication engineer, these two components are called *signal* and *noise*.

There are seven separate aspects to the time sampling problem:

(a) Purpose of the investigation.
(b) Length of observation period.
(c) Frequency of observations within the period.
(d) Averaging time of individual observations.
(e) Lag time of one variable with respect to another.
(f) Choice of the best index of time variability when an empirical method is used.
(g) Nonlinear interactions.

(a) The purpose of the investigation is relevant, of course. One scientist's "signal" is "noise" to another. In some micrometeorological experiments, as many as 100 measurements per second are required from a sensor. This would add undesirable clutter for an agrometeorologist seeking to relate corn yields with weather factors.

(b) The length of observation period should be a simple multiple of a well-established periodicity, e.g., of a day or a year. If only part of a cycle is present, the data exhibit *trend*, which becomes a needless complication. Unless the purpose of the investigation is to determine trend, every effort should be made to limit the data to periods of steady-state conditions. In engineering studies of wind and snow loadings on buildings, for example, a 100-year record of weather observations may yield misleading predictions if the local environment was once rural but has since become urban. Particularly in studies of extreme values, the choice of observation period is one of the most difficult problems in biometeorology.

(c) The frequency of observations within the period is dependent upon the type of problem. Whenever possible, however, a continuous record should be maintained; even though not all of the data may ultimately be analyzed, the investigator retains flexibility in choosing the segments of most interest, i.e., when atmospheric stresses are greatest. For example, if measurements are made only once or twice a week, the most interesting events may be missed, and it may not be possible to undertake a case-history study of the increase and decay of stress and response. Another example is the diffusion of pollution from a chimney when only one downwind sector is of interest; unless equipment is available that automatically activates the sensors when the wind is from that direction, it is preferable to record continuously.

(d) The choice of interval for averaging individual observations is dependent in part on the limitations of the sampling equipment (tree rings yield only yearly values of growth; lead peroxide candles give monthly values of SO_2). In some cases there is a physical basis for selecting the averaging time; Pasquill's diffusion model (16, p. 98) prescribes the averaging time for turbulence measurements to be used when estimating ground-level concentrations downwind from a chimney over open countryside. In other applications, determination of the most appropriate averaging time is an essential part of the investigation. One procedure that should be discouraged is the measurement of instantaneous values at fixed intervals of time, or averaged values that are themselves separated by fixed intervals of time (such as a 1- or 10-min wind speed once an hour). This introduces *aliasing*, a mathematical term that is illustrated in a simplified manner in Fig. 7 (17, p. 37). When six equally spaced values from a sine wave are plotted, it is not possible to discriminate between the one-cycle and four-cycle curves shown in the figure.

(e) In attempting to relate stress to response, it is important to recognize that a lag time may be present, and that it may be variable. In studying the relation between water levels on Lakes Michigan and Huron and monthly precipitation in the basin, Muller *et al.* (18) have found evidence for a lag of about one-quarter cycle for periods of up to about 3 years. When rainfall was above normal for 8 months, the highest water level was reached about 2 months after the rainfall peak; when rainfall was above normal for 16 months, the lag was about 4 months. Lag in this case must therefore be prescribed in fractions of a cycle rather than in terms of a fixed number of months. In many environmental investigations, time-averaging and time-lag difficulties completely obscure any connections that exist among several variables. Cospectrum analysis (Chapter 7) is one method of studying lagged responses.

(f) When an empirical approach is being used in a search for relations

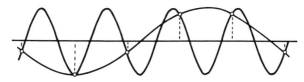

Fig. 7. Example of aliasing. In an analysis of observations taken at the times represented by the small open circles, dissimilar wavelengths cannot be distinguished.

between meteorological and biological variables, the choice of a single index to represent a stress such as low temperature is difficult. The average daily temperature or even a measure of its variability is meaningless when, for example, only frost occurrences are important; in other instances, a weighting system may be required if the rate of change of response is not constant with decreasing temperature, or if the duration of cold weather is significant. A *cold wave* is an important event physiologically. The phrase is meaningful to the layman but is impossible to define except in an arbitrary way. Chang (19) suggests that the term "cold wave" be limited to cases in which the temperature throughout the day is at least 10°C lower at any hour than 24 hr earlier, and that the minimum for the day be 0°C or lower. This 24-hr temperature drop is called the *interdiurnal decrease in temperature* (IDT), and its magnitude depends mainly on advection. Using this criterion, cold waves occur about 13 times a year in New York City; other arbitrary definitions would yield different frequencies.

Computers can calculate innumerable indices without difficulty but the real need is for insight into the nature of the problem, particularly the effect of physiological and human factors. Degree-days have often been used as an index for estimating fuel consumption in winter. In New York City, however, apartment dwellers use an almost constant fuel supply; when there is a thaw, windows are opened but thermostats are not turned down (B. Davidson, personal communication). This is an example of the human dimension that is often present in biometeorology.

(g) Most physical relations among three or more variables apply only to steady-state conditions (no trend), i.e., for periods of only an hour or so in most atmospheric applications. Suppose Eq. (2.1) is valid when hourly values of y, x_1, x_2, . . ., x_n are used.

$$(2.1) \qquad\qquad y = F(x_1, x_2, \ldots, x_n)$$

Then, to obtain a monthly average $\bar{y}$, the hourly values of y should be calculated from Eq. (2.1) and averaged. Because monthly mean values $\bar{x}_1, \bar{x}_2, \ldots, \bar{x}_n$ are sometimes readily available from climatological publications, however, these are sometimes substituted into Eq. (2.1) to obtain an estimate of $\bar{y}$. This practice should be discouraged because of the frequent occurrence of *nonlinear interactions*. As a simple example, wind chill increases with decreasing temperature or with increasing wind speed or both; because very low temperatures are usually associated with light winds, however, an estimate of wind chill derived from monthly mean data is not the same as that obtained from summing hourly estimates. Using the Siple formula (see Section 9.3), Court (20) illustrates the problem with the data in Table III; the average of the three

TABLE III
ILLUSTRATIVE EXAMPLE OF A NONLINEAR INTERACTION[a]

Temperature (°C)	−18	−34	−51
Wind (m/sec)	18	9	0
Wind chill (kcal/m²-hr)	1.8×10^3	2.1×10^3	0.8×10^3
Mean wind chill:		1.6×10^3	
Wind chill computed from mean temperature and wind:		2.1×10^3	

[a] Court (20).

wind-chill values is 1.6×10^3 kcal/m²-hr but when the mean temperature and wind for the three cases are used, the wind–chill is 2.1×10^3 kcal/m²-hr, or 34% more. As another example, Scorer (21) has pointed out that if the wind blows for half the time at 1.1 m/sec (2.5 mph) and half the time at 7.8 m/sec (17.5 mph), the average pollution from a point source is 2.5 times that accumulating in a steady wind of 4.5 m/sec (10 mph).

There is at least some physical basis for assuming that over a uniform surface when weather conditions are steady for an hour or so, the evaporation rate E varies as the product of wind speed u and the vertical vapor pressure gradient Δe (10, p. 93).

$$(2.2) \qquad E \propto u\,\Delta e$$

Using an overbar for the monthly mean value and a prime to indicate an hourly departure from the mean,

$$(2.3) \qquad \bar{E} \propto \langle u\,\Delta e \rangle_{\mathrm{av}} \propto \langle (\bar{u} + u')(\overline{\Delta e} + \Delta e') \rangle_{\mathrm{av}}$$

Thus, since

$$\overline{u'} = 0 \qquad \text{and} \qquad \overline{\Delta e'} = 0$$

$$(2.4) \qquad \bar{E} \propto \bar{u}\,\overline{\Delta e} + \langle u'\,\Delta e' \rangle_{\mathrm{av}}$$

The quantity $\langle u'\,\Delta e' \rangle_{\mathrm{av}}$ is not zero usually, because winds and vapor pressure gradients both have diurnal cycles and therefore are correlated.

2.2. CONCENTRATION AND DOSAGE

The concentration of a gas (or of suspended particulate matter) is expressed as mass per unit volume. Alternatively, concentration of a gas is given in parts per million, hundred million, or billion (ppm, pphm, or ppb) by volume. The two systems are related as follows:

$$(2.5) \qquad 1 \quad \mathrm{ppm} \times (\text{molecular weight} \times 10^6/22{,}400) = 1 \quad \mu\mathrm{g/m^3}$$

at standard temperature and pressure. Mass per unit volume is preferable because it is compatible with building ventilation rates (volume per unit time), chimney emission rates (mass per unit time), or vertical transfer rates of CO_2 and water vapor from vegetation (mass per unit surface area per unit time). The concentration of water vapor in mass per unit volume is called *absolute humidity*.

Concentration is sometimes a useful index, particularly when the response of the organism is almost instantaneous, as in the case of odor. The flow rate through the nose is large and the amount of odor-sensing tissue is small; when threshold concentrations are exceeded for as little as 1 sec, odor is detected (22). A meaningful but simple statistic in this case is, therefore, the percentage of time that the concentration is above the threshold.

Figure 8 is a schematic representation of the variation of concentration C of some gas with time t. The time scale is not labeled and could be a second or a year. The total area under the curve is defined as the *dosage D*, in units of mass-time per unit volume (or ppm-time).

$$(2.6) \qquad\qquad D = \int_{t_0}^{t_1} C \, dt = \bar{C} \, \Delta t$$

where $\bar{C}$ is the mean value of C over time interval $\Delta t = t_1 - t_0$. There are many applications in which dosage is more relevant than concentration, particularly when the response time to stress is long. Note that if a man's breathing rate is x cm³/hr and if he has been exposed to a dosage of y g-hr/cm³, then xy g of pollution have entered his lungs.

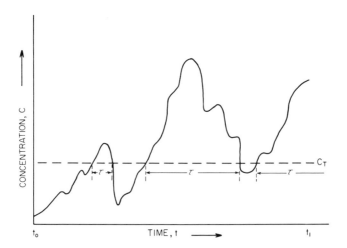

Fig. 8. Schematic representation of the variation of concentration C with time; C_T is the threshold concentration for some physiological response.

Dosage above a threshold value C_T is sometimes more useful.

$$(2.7) \qquad D_T = \int_{t_0}^{t_1} (C - C_T)\, dt, \qquad C \gtrless C_T$$

The severity of an odor problem may then be rated by the magnitude of D_T rather than by the percentage of time that the threshold concentration is exceeded.

Despite the widespread acceptance of the concept of dosage, there are cases, nevertheless, in which D and D_T do not provide sufficient information about the nature of atmospheric stress. Moderately high concentrations occurring for a long time may yield dosages (or values of D_T) of the same order of magnitude as extremely high concentrations lasting for only a short period. The effects on life are not necessarily the same, however. Dosage per unit time is therefore preferable sometimes as a normalization technique, i.e., values may be given for the quantity D_T/t, where t is the length of time that $C > C_T$.

The length of time τ that the concentration C remains continuously above the threshold C_T on any one occasion is a statistical variable called a *dosage event*. The number n of such events in a long record is another statistical variable. The quantity τ depends partly on the value of C_T. If the threshold is zero, the dosage event is the complete record and $n = 1$. On the other hand, if the threshold is never exceeded, $\tau = 0$ and $n = 0$. Between these two extremes, τ and n can assume many values. Harris and Dubey (23) have proposed a statistical prediction model for estimating the frequency distribution of dosages from a knowledge of the distributions of τ and n. An example of a computer tabulation of actual data, given by Larsen *et al.* (22) for a sampling station in Philadelphia, is shown in Table IV. The threshold concentration for SO_2 has been set at 0.24 ppm.

Air quality standards are often expressed as an average value of a pollutant over a given period of time. In the case of SO_2, a common standard is

0.3 ppm averaged over 8 hr (dosage of 2.4 ppm-hr)
1.0 ppm averaged over 1 hr (dosage of 1 ppm-hr)
1.5 ppm averaged over 3 min (dosage of 0.075 ppm-hr)

Results of the type shown in Table IV are useful in determining whether air quality standards have been met.

Other considerations of a different nature may be illustrated with examples from the field of corrosion chemistry. It is known that SO_2 in a moist atmosphere corrodes materials, and it is of engineering importance to be able to specify corrosion rates for a given metal and level of SO_2.

TABLE IV

DOSAGE AND DURATION OF SULFUR DIOXIDE EVENTS EXCEEDING 0.24 PPM IN PHILADELPHIA, DECEMBER 1, 1963–DECEMBER 1, 1964[a]

Dosage (ppm-hr)	Duration (min)									Total no. cases
	5	10	15	20–30	35–60	65–120	125–240	245–480	485–1440	
0.0000–0.0009	24	4	1	2	0	0	0	0	0	31
0.0010–0.0019	13	9	5	0	0	0	0	0	0	27
0.0020–0.0049	15	18	7	8	2	0	0	0	0	50
0.0050–0.0099	5	12	5	19	2	1	0	0	0	44
0.0100–0.0199	3	7	9	27	11	0	0	0	0	57
0.0200–0.0499	0	0	2	25	42	14	0	0	0	83
0.0500–0.0999	0	0	1	7	17	22	4	0	0	51
0.1000–0.1999	0	0	0	3	7	22	11	1	0	44
0.2000–0.4999	0	0	0	0	1	15	28	2	0	46
0.5000–0.9999	0	0	0	0	0	0	7	11	2	20
1.0000–1.9999	0	0	0	0	0	0	0	5	6	11
2.0000–4.9999	0	0	0	0	0	0	0	0	3	3
Total no. cases	60	50	30	91	82	74	50	19	11	467

Total dosage = 67.23 ppm-hr

Total duration = 586 hr, 55 min

[a] Larsen et al. (22).

However, a protective film of rust may form during the first few days of exposure, and this has a significant effect on the subsequent behavior of the sample. Therefore, initial conditions are important. Furthermore, iron continues to rust even if the damaged specimen is moved to a location containing no SO_2 but high relative humidity (24); there appears to be continuous regeneration of $FeSO_4$, one atom of sulfur producing many molecules of rust. Hence, neither concentration nor dosage might be a useful index. Metals are also corroded by salt (NaCl). In this case a weak solution produces a much larger galvanic electrolytic reaction than does a highly concentrated one, and the greatest damage to steel occurs with a solution of about 4% NaCl.

These examples emphasize the importance of determining experimentally the receptor response function. The receptor may be a single living organism. Alternatively, it may be a population, and the response function may be given conveniently as the frequency of complaints, the number of cases of eye irritation or legal action, the number of leaves or plants that have been damaged, etc.; Singer (25) uses a log-normal form for population response, the parameters in the equation being fitted from experimental data (Fig. 9). Many kinds of ecological response seem

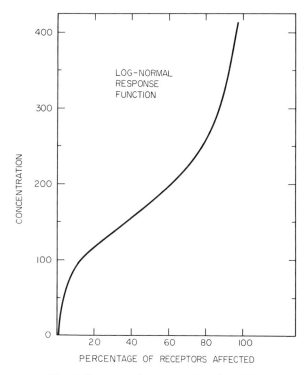

Fɪɢ. 9. Log-normal response function (25).

to be of the log-normal type (25, 26). Sometimes the curve is param-
eterized by a single number, TL50 (50% tolerance level), which is the
concentration that affects 50% of the population for a given exposure
time. Green (26) suggests that TL50 may be fitted to a hyperbola of
the form

$$(2.8) \qquad\qquad \text{TL50} = a + b/t$$

where a, b are empirical constants and t is the exposure time. The
quantity TL50 is then, for example, a concentration in parts per million
which approaches infinity as t approaches zero, and approaches a constant
value [a in Eq. (2.8)] for very long exposure times. In the case of
alfalfa injury by SO_2 in the laboratory, the parameters a, b have the
empirical values given in Table V (27, p. 421); the table includes
relevant results not only for TL50 but also for TL100 and for the case
of "incipient damage." To assist in rapid comprehension of the empirical
relations, Heck *et al.* (28) have employed a three-dimensional visual
display, as illustrated in Fig. 10, which concisely summarizes a large
amount of information. Separate graphs are required, of course, for

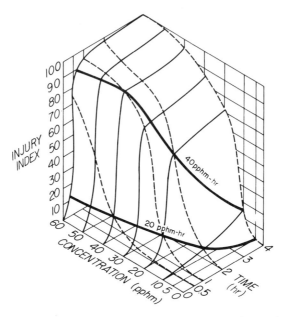

Fig. 10. Three-dimensional display of the interrelations of time and concentration on the sensitivity of pinto bean plants to ozone (28).

different species and for different types of pollution. In addition, reproducibility outdoors is difficult to achieve because the concentration of pollution is no longer constant.

Duckworth and Kupchanko (29) have introduced the concept of *dosage–area* as an urban pollution index, using examples from the San Francisco region. The severity of a smog day depends not only on the dosage but also on the size of the affected area. Assuming the existence of a satisfactory network of samplers, the first step in the analysis is to draw isopleths of equal dosage. Next, the areas within each annular ring are measured with a planimeter. The dosage–area for a particular

TABLE V

VALUES OF QUANTITIES *a*, *b* IN EQ. (2.8) THAT PRODUCE THE INDICATED DEGREE OF INJURY FROM SO_2 IN ALFALFA UNDER LABORATORY CONDITIONS[a]

	a	*b*
For incipient injury	0.24	0.94
For 50% leaf destruction	1.4	2.1
For 100% leaf destruction	2.6	3.2

[a] Brandt and Heck (27).

annular ring may then be obtained as the product of area and mean dosage. Finally, the annular values are summed to yield a metropolitan total.

Although the dosage–area index is not sensitive to a single peak concentration in time or space, it does highlight the regional severity of an "episode" and it is of value in comparing consecutive days during unchanged meteorological conditions. In one example (29), the dosage–area was higher on the second day in the San Francisco area than on the first, suggesting an accumulation of pollution in the basin. Csanady (30) has attempted to develop a statistical model of dosage–area, assuming that the quantity is a random variable.*

In summary, concentration and dosage are usually easy to measure or derive and may be of considerable value in certain investigations. In other cases, however, they may have little or no relation to receptor response. Some examples of this have been given while other aspects will be discussed in Sections 2.6–2.8. As a final illustration, measurements of absolute humidity, although useful in many meteorological situations, are irrelevant to the problems of forest fire hazard or of the rates of swelling of tree buds in spring; the proper moisture indicator then is *relative humidity* (the ratio of the actual vapor pressure to the saturation vapor pressure). (See also Question 36 in the Appendix.)

2.3. RADIOACTIVITY UNITS

An examination of radioactivity units is instructive. Human response to radiation is complex, and health physicists have made a serious attempt to normalize biological effect, culminating in a unit called the *rem.*

The activity of a radioactive substance is given by the number of disintegrations occurring in unit time. The unit usually used is the curie (c) which corresponds to 3.7×10^{10} disintegrations per second and is approximately the rate of disintegration of 1 g of radium.

The commonly occurring types of radiation are α-rays, β-rays, γ-rays, X-rays, and neutrons. Although differing markedly in physical properties, all give rise directly or indirectly to ionization in substances with which they interact and, for this reason, are often referred to collectively as ionizing radiation.

α-Rays are positively charged helium nuclei. They are characterized by dense ionization along their paths in matter, although the depth of

* See Question 15 in the Appendix regarding the area–depth curve in precipitation studies, a concept not entirely dissimilar from that of dosage–area.

penetration is limited and in water or tissue amounts to only about 0.1 mm.

β-Rays are positively or negatively charged electrons. Their range in tissue is typically of the order of a few centimeters.

γ-Rays, electromagnetic radiation, transfer energy to matter with which they interact by excitation or ejection of electrons (secondary radiation). The range of γ-rays in matter can be very large, although quantitative estimates are not meaningful. While extreme ranges in tissues are possible, the number of photons progressively decreases. In this case, a more useful concept is the thickness required to reduce the intensity by half.

X-rays are also electromagnetic radiation. They are usually produced by processes outside the nucleus rather than within it as is the case with γ-rays. X-rays are normally produced by decelerating high-speed electrons.

Neutrons are constituents of the nucleus and are uncharged particles. Because the behavior of neutrons and α-rays can be very complex, only X-, γ-, and β-rays will be discussed in this section.

The effect of ionizing radiation on matter depends both on the nature of the radiation and on its energy. The latter can be expressed in any of the conventional units, but is normally given as electron volts (eV) or some multiple thereof such as kiloelectron volts (keV) or mega electron volts (MeV). One million electron volts are equivalent to 1.6×10^{-6} ergs.

The effect of ionizing radiation on matter is related to the energy absorbed per unit mass—the *absorbed dose*—and is measured in *rads*. One rad is 100 ergs/g.

For many purposes, it is more convenient to use the *exposure*. The unit is the *roentgen* (r). One roentgen is an exposure of X- or γ-radiation such that the associated corpuscular emission per cubic centimeter of air at normal temperature and pressure produces, in air, ions carrying one electrostatic unit of quantity of electricity of either sign.

The relationship between the rad and the roentgen is complicated, depending not only on the material which is exposed but also on the photon energy. In air, 1 r corresponds to about 87 ergs/g or 0.87 rads. In water, 1 r corresponds to about 97 ergs/g or 0.97 rads at photon energies between 0.2 and 1 MeV but to about 85 ergs/g or 0.85 rads between 0.01 and 0.05 MeV.

Both the roentgen and the rad refer to total doses. Dose rates are given by the roentgen per unit time (*exposure rate*). The exposure rate from natural background is about 10 μr/hr, though there are variations with latitude and nature of the soil.

Measurement of radiation presents many difficulties, and great care must be exercised in the interpretation of measurements. The Geiger–Müller and scintillation counters measure the disintegration rate or radioactivity of a sample. Most instruments in common use require careful calibration. The responses of these instruments are strongly energy dependent, and calibration must be done using a standard of the same or at least of very similar energy. The geometry of the counting arrangement must be controlled, not only with respect to the relationships between the source and the instrument but also with respect to the distance, nature, and quantity of surrounding materials. Such expressions as curies of mixed β emitters or mixed $\beta\gamma$ emitters are virtually meaningless unless the composition of the mixture is precisely known and even then, because of differential decay rates of the components, the composition is time dependent.

Various devices are in use to measure exposures or exposure rates. Absolute measurements can only be made using highly refined equipment and techniques, and great care has to be exercised in the interpretation of measurements. Many devices are strongly energy dependent and are sensitive to geometrical relationships between source, absorber, and surroundings.

One other unit which is often encountered is the *rem*. When considering only the effects of X- or γ-radiation, the rem is for most purposes numerically equal to the rad. The use of the rem is related to the biological effects of neutrons and α-particles. It is sufficient to note here that experience shows that, rad for rad, the biological effect depends on the type of radiation. In the field of radiological protection, the term *quality factor* (QF) has been introduced. This is simply a modifying factor that permits normalization of biological effects produced by different types of radiation per unit of dose. The product of the dose in rads multiplied by the quality factor is the dose equivalent. The unit is the rem (1 rem = 1 rad $\times$ QF), and it normalizes for biological effect.

Biological effects are divided into two main classes, those which directly affect the individual exposed (*somatic effects*) and those which, through damage to the genetic material of the individual, may only become manifest in his progeny (*genetic effects*).

Somatic effects can be divided into two classes, depending on the time scale required for the effect to become manifest. Acute radiation sickness is a well characterized syndrome following doses of about 100 or more rads. At lower doses, recovery from the immediate symptoms is usually complete, but at doses exceeding 500 rads, death becomes increasingly probable. Among the survivors of acute radiation sickness as well as others receiving lower doses, the principal long-term effects are leukemia and cancers which may develop in periods measured in years following exposure. Nevertheless, the lowest dose which has been observed to produce leukemia in humans is about 100 rads, and this is still considerably higher than the levels normally found in the environment. It is not known whether or not a true threshold level exists for any or all the somatic effects of radiation below which complete recovery would occur. Nor is it known whether the risk of a given biological effect is linearly related to the dose. In practice, the assumption is usually made that no threshold exists, and linear dose–effect relationships apply. In any case then, it is apparent that doses of the order of tens of rads will be associated with an incidence of somatic effects comparable with the natural incidence of leukemias and cancers in human populations.

The biological response to radiation is extremely complex. Of the many sources of additional information, the reports of the United Nations Scientific Committee on the Effects of Atomic Radiation are recommended, being both authoritative and concise (31, 32).

2.4. DEGREE DAYS

Degree days are the accumulated departures of mean temperatures on successive days from some reference temperature (negative departures for space heating, positive departures for air conditioning and vegetation growth). If the mean temperature for a winter day is 3°C and the reference value is 15°C, then 12 degree days are added to the previous total. Similarly, *degree hours* are the accumulated departures of hourly temperatures from some reference value.

Whereas the time integral of concentration is dosage, the time integral of temperature is degree days; in fact, if the ordinate scale in Fig. 8 is relabeled as temperature, the area above (or below) the dashed line represents degree days or degree hours. Similarly also, normalization to unit time is often desirable; the degree days are commonly compared with the climatological mean value for the same date. This is satisfactory for winter space-heating requirements; in agricultural applications, however, a particular degree-day accumulation by late summer may be associated, for example, either with seasonable weather throughout the growing season or with generally cool temperatures compensated by one or two heat waves; the crop yields are different in the two cases.

The heating industry has adopted various reference temperatures in different countries, 15°C in Belgium and 18.3°C (65°F) in North America, for example. The initial date of the season is also established in an arbitrary way; in Belgium, it is the first day in September or October when the outdoor mean temperature is less than 15°C and concurrently the maximum is less than 18°C (33). However, the consumption of fuel in private homes is reasonably well correlated with any base temperature and initial date; thus, there is no reason why consumption per week could not be correlated directly with mean weekly temperature. In any case, the correlation is not perfect because the heat losses from a building depend on a large number of factors, e.g., wind speed, solar radiation, and thermal insulation.

A similar term is the *freezing degree day*, in which the reference temperature is taken as 0°C (34). This is used as an index of ice growth in arctic waters.

In agricultural applications, *degree days* are sometimes called *accumulated heat units* (AHU) (35). To provide a steady flow of fresh peas to canning factories in Britain, for example, plantings are made at equal intervals of AHU's, using a reference temperature of 4.4°C (40°F). In the case of milk production, on the other hand, degree hours above 27°C (80°F) have been used as an index; temperate-zone cattle often begin to show signs of impairment of their heat-regulating mechanisms above 27°C (36). As still another illustration, Davitaya (37) has used 10°C as a reference temperature in studies of the yields of sugar beet and of sunflower in the U.S.S.R.

Physiological response depends on a number of factors in addition to temperature. In regional comparisons of the growth rates of agricultural crops, for example, the number of hours of daylight is important. Nuttonson (38) has suggested the *photothermal unit* (PTU) (°C-hr), in which degree days are normalized to a standard length of day such as 10 hr. Equation (2.7) is modified as follows:

$$(2.9) \qquad \text{PTU} = \int_{t_0}^{t_1} (T - T_1)(H - H_1)\, dt$$

where T_1, H_1 are the reference temperature and number of hours of daylight, respectively, and $T > T_1°C$, $H > H_1$ hr.

For chemical reactions, the rate k is often given by an equation of the form

$$(2.10) \qquad k = a \exp(-b/T)$$

where a and b are positive constants. Thus, the reaction rate increases with increasing temperature; corrosion damage is negligible in the deep freeze of winter, for example. A common way of expressing the variation of k with temperature is by use of the ratio R, which is an index of the effect of a 10° rise in temperature;

$$(2.11) \qquad R = k(T + 10)/k(T) = \exp[-b/(T + 10)]/\exp(-b/T)$$

Lee and Sharpe (39) suggest that Eq. (2.11) is sometimes applicable to biological data, provided that the temperature range is relatively small. Then the ratio R, called a *temperature quotient*, may be a useful environmental index.

The CO_2 respiration rates also depend on temperature, as do photosynthetic rates. In the latter case, climatic adaptation has been shown to be important; the optimum temperature is about $12°C$ in alpine regions, $20°C$ in temperate zones, and $30°C$ in the tropics.

Yevjevich (40) suggests an approach to the study of drought that is conceptually very similar to that of dosage and degree day. Instead of concentration or temperature, the variable in this case is water deficiency, using as index the difference between precipitation and evaporation. An arbitrary threshold value is chosen, perhaps a zero water deficiency. Then the cumulative deficiency and the time between threshold crossings are variables that may be examined statistically.

2.5. CONCENTRATION AND FLUX

The *flux* of a quantity is its rate of transfer through a medium. Flux of matter (water vapor, CO_2, O_3, etc.) is usually expressed in grams per square centimeter per unit time. Heat flux is in calories per square centimeter per unit time, i.e., ly per unit time, where a *langley* (ly) $= 1$ cal/cm^2.

Fluxes are rarely measured directly but are inferred from concentration or temperature changes in space or time. Suppose that CO_2 concentrations C_1 and C_2 (averages for an hour or so) are measured at two heights z_1 and z_2 $(z_2 > z_1)$ above a large uniform field of wheat. If $C_1 = C_2$, the flux is zero; if $C_1 > C_2$ (see Fig. 11), the flux is upward, i.e., the vertical *gradient* of concentration may be used to decide the direction of flux and is one factor in estimating its magnitude.

If the measured concentrations show a trend towards higher (or lower) values during the sampling period, and if the wheat field is so large that the wind is not bringing CO_2-enriched air from another area, then there must be vertical *flux divergence*. Thus, if CO_2 is increasing in the layer between the samplers, the upward flux at level z_1 must be greater than the upward flux at level z_2 (see Fig. 11). Similar arguments apply when the flux is downward and/or the concentrations are decreasing in time.

Flux, of course, is a vector quantity and in many cases must be separated into three orthogonal components. Horizontal as well as vertical fluxes and flux divergences might be present, if the wheat field were small.

Biological response in many cases depends on the flux at the surface

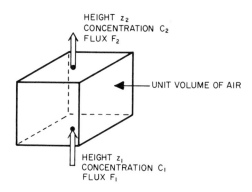

HEIGHT z_2
CONCENTRATION C_2
FLUX F_2

— UNIT VOLUME OF AIR

HEIGHT z_1
CONCENTRATION C_1
FLUX F_1

FIG. 11. Schematic representation of concentrations and vertical fluxes over a uniform surface. If $C_1 > C_2$, there is a gradient, and the flux is upward. If $F_1 > F_2$, there is flux divergence, and the value of C is increasing within the unit volume of air.

of interest, *not* on the concentration or dosage in the air. This fact has not always been recognized and, as a result, some experimental investigations have been rather meaningless. In studies of heat stress on a body, for example, the surface heat flux is more important than air temperature (see Section 8.6). Admittedly, a partial correlation may exist between air temperature and surface heat flux, which in turn may produce a correlation with heat stress, but the heat flux to or from the body is often the fundamental variable. Similarly, one would not expect to find a relation between evaporation from the ocean and the humidity at a given level, or between the CO_2 exchange rate from vegetation and the concentration of CO_2 in the air. On the contrary, gradients of humidity and CO_2 are required.*

In the case of a pollution stress, few attempts have been made to distinguish between concentration and flux. If the surface of the earth were a perfect reflector for gases, aerosols, and suspended particulates, as most diffusion models assume, there would be little damage to vegetation. In actual fact, the transfer rates at the leaves themselves are of fundamental concern (see Section 8.7).

One of the isolated studies suggesting the importance of flux is the

* This line of reasoning can be extended one step farther; flux divergence may be more important than flux in some cases. Solar radiation has been correlated with agricultural yield, but radiative flux divergence through the canopy may be a better index.

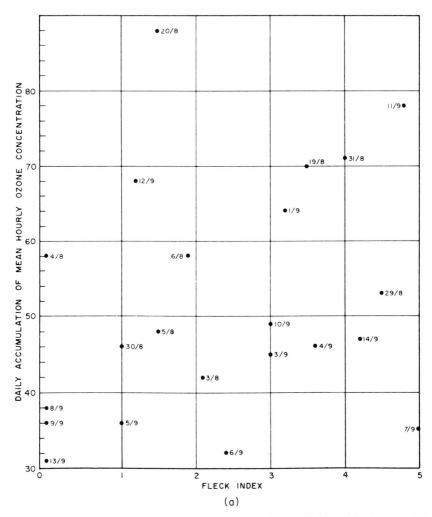

Fig. 12. Scatter diagrams relating the index of tobacco flecking (abscissa) to (a) daily oxidant dosage, and to (b) a quantity roughly proportional to oxidant flux. Measurements made in 1961 at Port Burwell, Ontario (41). Numbers beside points indicate date of month.

work by Mukammal (41, 42) on the flecking of tobacco leaves in fields along the north shore of Lake Erie. The leaf damage was similar to that obtained in laboratory fumigations with ozone. Yet daily measurements of oxidant dosage outdoors (believed to be mainly ozone) in the summers of 1960 and 1961 bore little relation to incidents of flecking; the wide scatter is shown in Fig. 12a. When the daylight oxidant dosage was

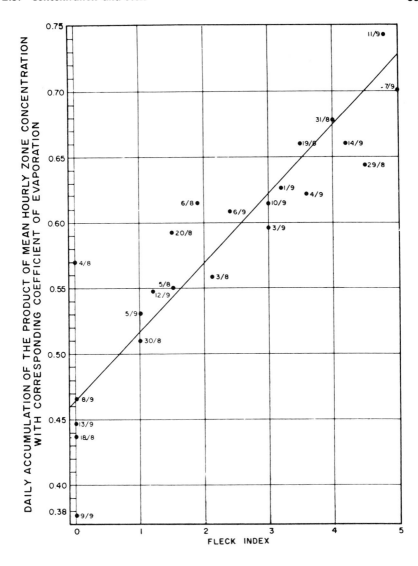

multiplied by a coefficient A, however, most of the scatter disappeared (Fig. 12b). The explanation seems to be that the new variable on the ordinate scale is nearly proportional to oxidant flux; the quantity A is simply the coefficient of evapotranspiration which is assumed proportional to that for oxidants. A few measurements of oxidant at two heights showed that a vertical gradient did indeed exist. In laboratory fumigation chambers, on the other hand, ozone is generated continuously, and the airflow is usually from ceiling to floor; thus, there is no concentration gradient above the leaves.

Concentration gradients are often slight and difficult to measure. Although meteorologists recognize this problem and have developed, for example, matched thermometers capable of detecting very small temperature and humidity differences, atmospheric chemists have been more interested in the absolute accuracy of one instrument than in the relative accuracy of a pair of sensors, which is a separate problem.

2.6. PHYSICAL TIME LAGS

Before commencing a discussion of the response of biological systems to stress (Section 2.7), some physical analogies may be helpful. As an introductory example, suppose that a mercury-in-glass thermometer at temperature T_0 is suddenly immersed into an air stream of temperature T_1 $(T_1 > T_0)$. The thermometer does not achieve instantaneous equilibrium with its new environment but approaches it asymptotically. The lag depends upon the physical properties of the thermometer and the flow rate of the air stream.

Apart from an initial unsteadiness when heat is being transferred through the glass, but not into the mercury, the thermometer response may be represented by the equation

$$(2.12) \qquad\qquad dT/dt = (T_1 - T)/\lambda$$

where

dT/dt = the rate of change with time of the indicated temperature,

λ = a constant of proportionality (dimension of time).

Equation (2.12) may be solved to yield

$$(2.13) \qquad\qquad (T_1 - T)/(T_1 - T_0) = \exp(-t/\lambda)$$

when the instant of immersion into the new environment is taken as $t = 0$.

For time $t = \lambda$, the right-hand side of Eq. (2.13) has the value 0.368, i.e., after λ sec in the new environment, the temperature difference $(T_1 - T)$ has been reduced to 36.8% of its original difference $(T_1 - T_0)$. The thermometer has then responded to 63.2% of the initial temperature difference. The quantity λ is known as the *time constant*. Equations (2.12) and (2.13) are said to represent a *first-order response*, i.e., Eq. (2.12) is a first-order differential equation.

The dependence of λ on wind speed and on the diameter of a thermometer is illustrated in Fig. 13 (43). As the wind increases or a thinner thermometer is used, the response becomes more rapid. The time constants in Fig. 13 are only fractions of a second whereas that of an ordinary mercury-in-glass thermometer is about 2–3 min.

Equation (2.13) may be used to determine the percentage responses of the sensor after different time intervals. The results are given in Table VI (44). To achieve 95% recovery, for example, an elapsed time of 3 λ is required.

TABLE VI

RECOVERY IN TERMS OF THE TIME CONSTANT λ OF A SENSOR OBEYING EQ. (2.13) AS A RESULT OF A STEP-FUNCTION INPUT[a]

Recovery, %	50	63.2	90	95	99	99.5
Elapsed time, λ	0.7	1	2.3	3	4.6	5.3

[a] Gill (44).

There is a conceptual similarity between the terms, "time constant of a thermometer" and "disintegration rate of a radioactive source." The latter depends only on the number of unstable atoms present at any given time; therefore, as successive disintegrations progressively remove the unstable atoms, the activity of a source *decays* exponentially with a decay constant, Γ per unit of time. Equal numbers of atoms of different radioactive substances do not disintegrate at the same rate and Γ may be in units of reciprocal seconds, minutes, hours, days, or years, depending on convenience. Another measure of the decay rate is the *half-life* $(t_{1/2})$, being the time for half the atoms to disintegrate. The half-life is given by the relationship, $t_{1/2} = 0.693/\Gamma$, which may be derived from an equation similar to Eq. (2.13), i.e.,

$$\exp(-\Gamma t) = \tfrac{1}{2}$$

In many cases, the product of disintegration is itself an unstable atom in which case the time course of the rate of disintegration may be quite complex.

The effect of a sinusoidal temperature wave is illustrated in Fig. 14. As the time constant increases, the response is increasingly damped and lagged; the thermometer is said to *filter* the original signal. The *frequency response* $R(f)$ of a filter is defined as the ratio of the indicated amplitude X of an oscillation of a given frequency f to the true amplitude X_0. In the example in Fig. 14, the amplitude ratio X/X_0 ranges from 0.99 for a thermistor to 0.28 for a large bare mercury-in-glass bulb.

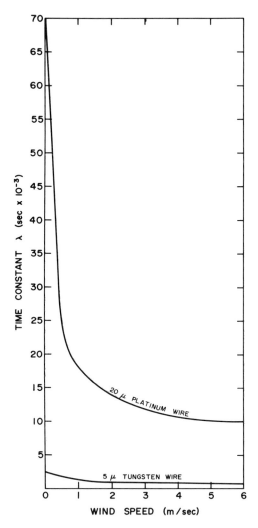

Fɪɢ. 13. Time constant of a resistance thermometer versus wind speed for two wire diameters (43).

The frequency response of a single filter to a sinusoidally fluctuating input is given by Eq. (2.14) (45, p. 151).

$$(2.14) \qquad R(f) = X/X_0 = (1 + 4\pi^2 f^2 \lambda^2)^{-1/2}$$

The phase lag ϕ $(0 < \phi < \pi/2)$ is given by (45, p. 152)

$$(2.15) \qquad \phi = \tan^{-1}(-2\pi f \lambda)$$

Some results are shown graphically in Fig. 15, as suggested by Gill (44). The ordinate is the ratio of indicated/true amplitude. The abscissa

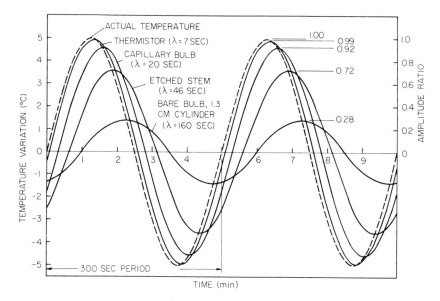

FIG. 14. Response of different temperature sensors to an environmental sinusoidal temperature cycle of 300-sec period (44).

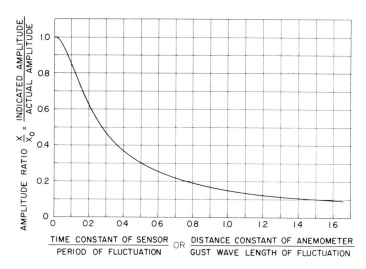

FIG. 15. Ratio of indicated to true amplitude of a sinusoidal fluctuation versus ratio of time constant of a thermometer to fluctuation period, or versus ratio of distance constant of an anemometer to the gust wavelength (44).

represents not only the time-constant/period ratio for a thermometer but also, in the case of an anemometer, the ratio of *distance constant* to the wavelength of a sinusoidal wind-speed fluctuation; the *distance constant* is the length of air column passing an anemometer propeller (initially at rest) required for the sensor to achieve 63.2% of the free-stream velocity.

A large number of physical sensors are characterized by the first-order response of Eq. (2.13). However, experimental verification is required in each case. Larsen *et al.* (46), for example, have shown that the time constant of a particular SO_2 sensor is about three times greater for increasing than that for decreasing concentrations.

Equation (2.12) can be used to improve the estimate of the true behavior of a meteorological or air quality variable when using a first-order response sensor. Rearranging the equation and introducing a more general variable y in place of temperature T,

(2.16) $$y_1 = y + \lambda \, dy/dt$$

where y_1 and y are the true and indicated values of the variable. The degree to which the true signal may be recovered depends upon the magnitude of the time constant. If λ is large, the rate of change of y with time may be too small to be measured. As noted by Panofsky and Brier (45, p. 153), "perfect reversal of the original smoothing is im-

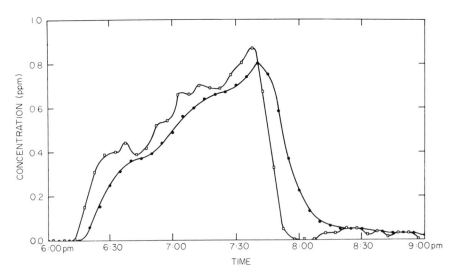

Fig. 16. Concentration of NO in Philadelphia on October 18, 1963. (□) observed concentrations using an instrument with a time constant of 10 min; (●) the corrected values using Eq. (2.16) (46).

possible because a wave which had previously been completely smoothed out of the series can never be recovered." Panofsky and Brier recommend time intervals of $\frac{1}{4}\lambda$ or less, when Eq. (2.16) is converted to finite-difference form.

After examination of a number of pollution samplers, Larsen et al. (46) suggest that a system be chosen with a time constant equal to about half of the shortest desired averaging time. An instrument with a time constant of 10 min should only be used for calculating means of 20 min or longer. Larsen et al. have developed a computer program for solving Eq. (2.16). An example, given in Fig. 16, shows the NO concentrations in Philadelphia on October 18, 1963; the time constant of the sensor was 10 min, and the recorded 10-min mean concentration underestimated the true value at times by 10%.

Not all sensors respond to an environmental change in the way indicated in Fig. 15. Sometimes a second-order differential equation is required to represent the balance of forces, yielding a solution that is called a *second-order response*. Examples include instruments with springs or floats, and first-order sensors attached to electric meters. The mathematical details will not be given here, but the net result is that vibrations may be induced that were not present in the original signal. Suppose, for example, that a galvanometer circuit is suddenly closed; the indicator needle overshoots, reverses direction, and again overshoots, executing thereafter a simple harmonic motion of decreasing amplitude (44).

When the input signal to a second-order sensor is fluctuating sinusoidally, there is the further problem of resonance, with indicated amplitudes being much larger than the true ones. Many wind vanes, for example, may overshoot in a turbulent wind, a problem whose solution has been given by Gill (44).

Many physical systems have a first- or second-order response to environmental change. The temperature inside an unheated building, for example, lags behind the value outdoors and the amplitude of fluctuations is damped. The time constant in this case is measured in hours and depends on wind speed, solar radiation, and the physical characteristics of the structure such as size, thermal insulation, and window area. The time constant in fact integrates the effect of all these variables, although its usefulness has not been widely recognized by the construction industry (perhaps because it cannot be determined readily at the design stage).

Takakura (47) has considered methods of predicting air temperature variation inside a glass house at night, taking into account radiative exchanges. In terms of the ratios A_f/A_g and V/A_g, where A_f, A_g are floor and glass areas and V is volume, his model predicts that:

(a) The phase lag of inside to outside air temperature increases with increasing A_f/A_g.

(b) The amplitude of inside air temperature variation decreases with increasing A_f/A_g and decreasing V/A_f.

On a smaller scale, Bryant (48) has investigated the effect of a Stevenson screen on the time constant of a thermometer. He determined experimentally in a wind tunnel that as the wind speed increases from 0 to 7 m/sec, the lag time decreases from 30 to 6.5 min. Bryant notes that these values are much larger than the time constant of an unshielded thermometer, suggesting that the screen has a dominating influence.

Examples of second-order responses include the resonance of structures in gusty winds and the development of oscillations (the *seiche*) in lakes. Second-order response problems present many difficulties but in all cases a fundamental parameter is the *natural period of vibration*, i.e., the period for which the system oscillates freely in the absence of external forces. As an indication of the order-of-magnitude of this quantity, Crawford and Ward (49) have reported that the natural period of vibration of a 19-story building in Ottawa was about 1 sec.

2.7. BIOLOGICAL CLOCKS AND TIME LAGS

Biological indices display both random fluctuations and regular time cycles; examples are given in Fig. 17 (50). Although the cycles are often in phase with geophysical periodicities, particularly with the diurnal and annual ones, the biological clock may continue to run when the living organism is transferred to an artificial environment. Constant-tempera-

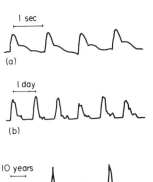

FIG. 17. Biological cycles, originating in (a) a small cell unit (human pulse), (b) in an organism (activity of a finch), and (c) in a population (numbers of lynx in Canada), respectively (50).

ture chambers with continuous illumination or darkness, or with an unnatural environmental cycle of day and night, have been used to reveal some curious effects. In general, the phase can be shifted easily but the 24-hr biological periodicity is much more difficult to alter. Not only does a seedling "remember" its natural environment but, even if it has never known the real world, it maintains 24-hr rhythms of growth in a darkened room (51, p. 257). Living things apparently inherit biological clocks from their parents, although Davis (52) does not exclude the possibility that the rhythms are caused by factors as yet undetermined.

A distinction is made between an environmental cycle (called a *Zeitgeber*) and a physiological cycle occurring within the organism (called a *circa rhythm*). The prefix "circa" indicates that the cycle does not always correspond exactly with a Zeitgeber (50). The 24-hr or *circadian* rhythm has been most widely studied but there is evidence for a circa-annual cycle (particularly in vegetation where a period of dormancy and/or chilling, called *vernalization*, is required), a *circa-tidal cycle* in some seaweeds, and a *circa-weekly* rhythm in man (due to a sociological Zeitgeber). An important consideration in the design of biometeorological experiments is the fact that external stresses may elicit different responses if applied at different points in a circadian cycle.

Aschoff (50) describes an experiment in which a man was placed in a soundproof underground bunker without time cues for 25 days. In the first 3 days, the volunteer maintained his relation with zone time; this was followed by 2 very long days and 1 extremely short one; thereafter, the man adopted a rather steady 26-hr cycle of activity and rest.

There are many indicators of circadian rhythms, such as body temperature, metabolism, cortisol in the blood, and excretion of potassium. When the sleep habits of man are disturbed by shift work or by air travel to distant locations, the individual rhythms may not remain in phase with each other or with the Zeitgeber. In fact, the interaction between a circadian rhythm and a changed Zeitgeber is of considerable physiological significance. Aschoff (50) emphasizes that a modest environmental stress may sometimes be severe in such cases.

Another interesting topic is the biological response to latitude changes. Flowering and fruiting are often critically dependent on the length of daylight, or *photoperiod*, and long-day plants will not flower or reproduce in the tropics, for example. Fogg (51, pp. 261–273) has given a useful introductory survey of the effects of modifying photoperiods in growth chambers; the number of hours of illumination, not the intensity of the light, seems to be the critical factor in the initiation of flowering and fruiting.

Finally, the determination of biological response times to environmental stress is a major consideration in many investigations. Curry (53) re-marks that if summer precipitation is in the form of occasional showers, the reaction of short-rooted grass is much more direct than that of a deep-rooted tree; the grass is crying "change" when the tree is hardly affected.

Plant physiologists have begun studies of response times. Evans (54, p. 243), for example, has summarized the results of experiments in which leaf temperature was measured following a sudden change in insolation. Time constants were about 10–40 sec for pepper and beans, 1–2 min for tobacco, strawberry, and cotton, and 5–20 min for succulents. In the case of photosynthesis, however, leaves of several species required up to 4 min to respond to an increase in light intensity but less than 30 sec to react to a decrease. Stark (55) has studied the transpiration rate of a small sagebrush plant during a controlled alternation of sun and shade. The results, shown in Fig. 18, reveal that the original transpiration rate was not restored when the shading was removed. Individual studies such

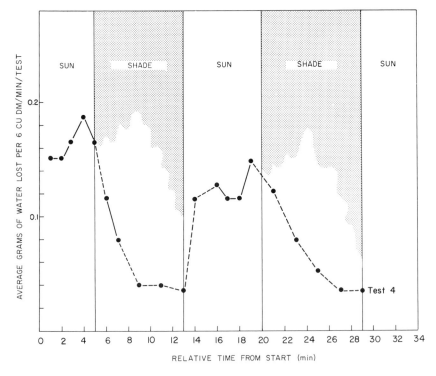

Fig. 18. Sun and shade transpiration rates for sagebrush, April 22, 1966 at 74°F or 23°C (55).

as these provide clues but not the broad outline of stress-response time lags.

There is little documentation of second-order biological response, although Fremlin (56) has given an ecological example. An animal population is presumed to be living undisturbed on an island. The abundance of food is assumed to control fertility, and the life span of the animal is long in comparison with the renewal rate of food. The animal population then may "overshoot" an equilibrium condition in both directions. At one stage, the island contains more animals than can be supported. Food production and reproduction rates decline, with a further population decrease due to starvation or death by old age. Eventually, the food supply is just sufficient for the population. However, old age further reduces the number of consumers, permitting the rate of food production to increase again.

On a much shorter time scale, the data of Parkhurst and Gates (57) show an oscillatory behavior in both surface temperature and transpiration rate of a cottonwood leaf during a 3-hr period; the experiment was performed under constant laboratory conditions. The leaf may have been near the upper limit of its "comfort zone," producing an alternating cycle of stomatal openings and closings.

In conclusion, the time-sampling problem is critical in environmental studies. The natural geophysical and biological rhythms are intertwined, and both are immersed in "noise." It is often asserted that an additional significant figure can be obtained in a result by merely averaging large samples of data. This is true if the observations are drawn randomly from the same population. In the time domain, however, there is almost always difficulty in deciding whether individual data points are in fact members of the same population, particularly when the stress-response rhythms and time lags are not well defined.

The analysis of physical and biological time series is very important and another entire chapter (Chapter 7) is devoted to the subject.

3/SAMPLING OF THE ATMOSPHERE: SPACE CONSIDERATIONS

3.1. SPACE VARIABILITY IN THE ATMOSPHERE

The *biosphere* is the transition zone between earth and atmosphere, within which most life is found. The meteorological elements in this region display great spatial variability in both the horizontal and vertical directions. Nevertheless, the gradients can usually be explained physically, when the appropriate band-pass filters are applied.

Macrometeorology is the study of large-scale weather processes, using surface observations at sites 200 km or more apart, and upper-air observations at even greater separation distances. Meaningful weather maps can be drawn for *synoptic* data (simultaneous observations, averaged over 1 min to 1 hr, from a network of stations), and the patterns are reproducible by two meteorologists working independently. When the observing points are 200 km apart, one sometimes wonders that isopleths can be drawn at all. However, small-scale noise is damped deliberately by careful selection of observing sites. The instruments are mounted at fixed heights over short grass, the sensors have similar response characteristics, and the thermometers are placed in standard Stevenson screens. The term *open exposure* indicates an absence of local obstructions such as buildings and trees.

Uniformity of siting is a disadvantage as well as an advantage for the biometeorologist. Historical records preserved by National Weather Services are not representative of the climate in a forest, in a garden, along the south wall of a house, or among the blades of grass in a lawn.

Similarly, annual frost-free periods vary across cities and across individual farms, because of local differences in the surface heat balance. In fact, isopleths of meteorological elements drawn on regional maps represent conditions over only a small fraction of the land surface; the lines merely indicate the values that would have been observed if trees and buildings had been replaced by short-grass surfaces near any point of interest.

On the *mesoscale* (a few to about 100 km), local topography has a predominant influence. An experienced mesometeorologist can estimate the medium-scale climatic features from a knowledge of the nature of the underlying surface, i.e., the orography, thermal properties, moisture availability and vegetative cover. *Topoclimatology* is the study of mesoscale climate in hilly or mountainous country.

On the *microscale* (less than ½ km), the standard weather-observing stations usually do not provide sufficient information, and additional sensors must be used at different heights above the ground and at "representative" nonrepresentative locations. Where local geometry is complex, as around a group of buildings or in a forest, the spatial variability creates problems in the planning of environmental experiments.

TABLE VII

CLASSIFICATION OF SPACE AND TIME SCALES

Motion	Rossby number[a]	Typical space scale	Typical time scale	Network spacing
Microscale	—	1 m	sec–min	cm–m
Gravity waves	10^2	$\lesssim 1$ km	min–hr	m–km
Mesoscale	10^2 to 10^1	5–10 km	hr	5–10 km
Small synoptic	1	100 km	hr–day	100 km
Large synoptic	10^{-1}	100–1000 km	days	100–500 km
Planetary	$<10^{-1}$	>1000 km	days–weeks	500 km

[a] The Rossby number is irrelevant on the microscale.

A classification of scales of motion is arbitrary and depends on individual preference or convenience. Nevertheless, Table VII may be a helpful qualitative summary. The *Rossby number* Ro has been used as one criterion.

$$(3.1) \qquad\qquad \mathrm{Ro} = V/fL$$

where V and L are characteristic velocities and lengths, and f is the Coriolis parameter, derived for each latitude from the component of the earth's rotation in the direction of the local vertical. On the microscale, f and Ro become irrelevant.

3.2. MICROMETEOROLOGY

Micrometeorological models and methods are closely linked to those in fluid mechanics. Problems of common interest include:

(a) Turbulent flow over a uniform surface of infinite extent.

(b) The modification of turbulent flow resulting from a discrete change in roughness, temperature, or moisture condition of the surface.

(c) Turbulent flow around spheres, cylinders, cubes, and other obstacles.

(d) Transfers of heat and water vapor at interfaces.

The wind tunnel is a useful experimental tool in fluid mechanics. Because uniform surfaces of infinite extent cannot be studied in the laboratory, however, there has been a search for very homogeneous outdoor sites, and field expeditions have been undertaken in Antarctica, Peru, the U.S.S.R., and Australia. It now seems clear that the main problems in micrometeorology have been solved for flow over ideal flat surfaces.

When there is a discrete change in the properties of the interface, an *internal boundary layer* (10, pp. 107–115) develops and deepens downwind. The air above this layer is not yet affected by the underlying surface. This kind of problem has its atmospheric counterpart not only on the smallest scale (flat-plate samplers for ragweed pollen, Section 3.6) but also in the flow of air from prairie to forest, land to ocean, and country to city.

Atmospheric flow frequently is disturbed by obstacles. The interference patterns depend on local geometry, wind speed, and wind direction (angle of attack). In ideal cases, von Kármán vortices (organized swirls of air) are shed at regular intervals in the wake of bluff bodies; however, the turbulence and air flow at street level in built-up areas are too complicated to be predicted from first principles.

Mrs. Ryd (58) refers to the *climatological sheath* around a building. The phrase is appropriate, indicating that the temperature, humidity, radiation balance, precipitation, and wind are all influenced by the structure. The temperature in a Stevenson screen over a grassy plot is often quite different from that in a ventilation and cooling intake, because of a recent trajectory in the latter case over the roof, or over an adjacent parking lot.* Micrometeorologists have been reluctant to study the climatological sheath because of the problems of reproducibility.

Buildings have a major effect on rainfall (see Section 3.6) and on snow depth patterns. Figure 19 shows isopleths of snowfall in two

---

* Variations in temperature are not too important provided that the wet-bulb value is constant.

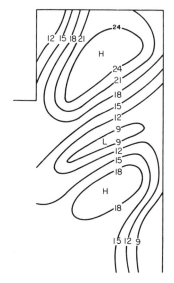
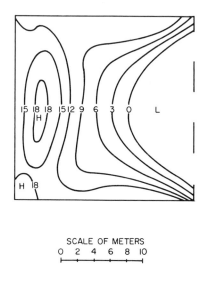

SCALE OF METERS
0   2   4   6   8   10

FIG. 19. Snowfall in centimeters in a courtyard (right) and in a partially enclosed court (left) in Montreal, December 28, 1964; ground was bare prior to the storm; snowfall in an adjacent open football field was a uniform 10 cm (P. Denison, personal communication).

adjacent courtyards in Montreal (P. Denison, personal communication). Similar variability is to be found in forest clearings and even over relatively open countryside, where much of the land remains wind–swept while large amounts of snow accumulate near fences and other obstructions. This may produce substantial local variations in the magnitude of the spring soil moisture recharge and in the date of commencement of vegetative growth.

The micrometerological gradients of temperature, humidity, and wind are largest close to the interface. When the sun is shining, the air within a few millimeters of the ground can be as much as 10°C warmer than that at a height of 3 m. On clear windless nights, on the other hand, a reverse gradient of equal magnitude can exist. Sutton (59) has said that "a seedling pushing its first leaf through the soil may, even in a temperate latitude experience in a single day a range of temperature that for man and other forms of life would correspond to daily commuting between the Tropics and the Arctic." Gates (60) uses the term *teleoclimate* to denote conditions at the surface itself.

The monitoring of meteorological elements in a vegetative cover presents special difficulties. Mukammal and Baker have described the horizontal spatial variations that exist on sunny days within a forest. One of their results is shown in Fig. 20 (61), a comparison of the diurnal cycle

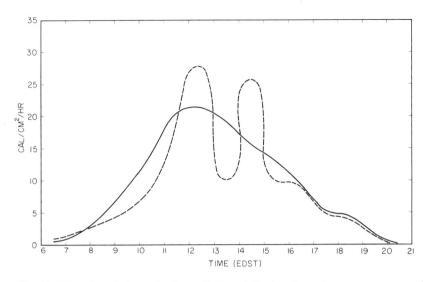

Fɪɢ. 20. Diurnal variations of solar radiation at fixed and moving sensors, averaged over two consecutive clear days in June beneath a pine canopy at Petawawa, Ontario (61): (——) 1.8-m solar radiation at moving sensor; (---) 1.8-m solar radiation at fixed sensor.

of solar radiation beneath a pine canopy as obtained from a fixed sensor and from another sensor that moved back and forth along a 30-m track at a rate of 6 m/min; the two instruments were at the same height. The hourly differences were sometimes large but surprisingly enough, the daily totals differed by only 3%.

On a smaller scale, large temperature gradients exist in the air space between adjacent sunlit leaves; indeed when winds are light, local circulation cells may develop.

Micrometeorology includes *cryptometeorology,* the study of enclosed spaces such as buildings, greenhouses, and caves. The biometeorologist is also interested in the heat balance of an animal or a human, and the effect of clothing. These questions require many new experimental data within climatological sheaths, including measurements of micrometeorological gradients in directions at right angles to sunlit walls and other nonhorizontal surfaces.

3.3. MESOMETEOROLOGY

There are four general classes of mesometeorological studies:

(a) Severe-weather analysis of squall lines, pressure jumps, hail storms, and tornadoes.

(b) Forecasting of airport visibility and cloud base for periods of 1–3 hr from local networks of stations.

(c) Mesoscale studies of precipitation, evaporation and clouds.

(d) Analysis of local circulations such as sea breezes and mountain winds.

For many years, mesometeorology was not given sufficient attention, partly because of preoccupation with the macroscale. However, many new tools have become available in the last several decades, including radar, tall towers, meteorological research aircraft, and automatic weather stations. Even satellites have been useful; for example, recent photographs have revealed the presence of von Kármán vortices, made visible by cloud patterns, downwind from the Madeira and Canary Islands; calculations show that the wind-tunnel theory applies on this much larger scale (62, 63). Paralleling the increased capability to probe the mesoscale, there have been comparable advances in computer technology. The expanding data streams have been assimilated, and old problems can be reexamined, including those that once required laborious hand calculations. For example, computers have produced clear-sky daily solar radiation charts for mountainous areas, given only the aspects of the slopes, latitude, and time of year.

The biometeorologist views a mesoscale phenomenon as part of an ecosystem. The physical structure of the system must be understood, but of equal importance is the effect of the mesofeature on vegetation patterns, on the behavior of animal and human populations, and on watershed management. The Chinook wind is believed to influence vegetation in Alberta (64) while the mountain wind at Denver, Colorado controls the ebb and flow of pollution over the urban area.

The physical meteorologist seeks local circulations under ideal conditions, when the macroscale winds and the topography are the same as postulated in his simplified models. For example, a straight coastline is likely to be chosen in experimental probing of the sea breeze. However, the lake-shore power plant is rarely built at an "ideal" site and furthermore, it emits pollution not only on days when a classical lake breeze is blowing but on every other occasion. Similarly, evapotranspiration takes place in watersheds when mesoscale circulations are poorly developed as well as when they are strong. In addition, the orography may be complex, as illustrated in Fig. 21 by a photograph of the Marmot Creek experimental basin on the east slopes of the Canadian Rockies (65). The estimation of precipitation and evapotranspiration is difficult in such areas.

The interaction between mesoscale systems is receiving increasing attention. Smith (66) has studied the nighttime valley wind at Johns-

FIG. 21. Aerial view of the Marmot Creek watershed in Alberta looking west (65).

town, Pennsylvania; the normal drainage patterns are disrupted in the vicinity of an area of heavy industry, and the air converges towards this "heat island" from all directions.

Another mesoscale problem concerns the estimation of areal precipitation. Landsberg (67) points out that if there is one rain gauge per 259 km² (100 miles²), only 10^{-10} of the area is being sampled. Information on mesoscale precipitation variations is urgently needed for both watershed and ecological analysis. The principal cause of spatial differences is orography. Figure 22, for example, shows large gradients of mean annual precipitation from south to north across Vancouver, B. C., associated with a rise in elevation of 900 m (68).

In planning winter recreational areas, a knowledge of mesoscale snowfall patterns is needed, including the year-to-year variability and the timing of spring snowmelt. One factor that must be recognized is that when the forest is cut, the mesoclimate will change. Historical data therefore require careful interpretation.

3.4. MACROMETEOROLOGY

Individual weather systems on the synoptic scale are well organized in space and time, and they can be followed on consecutive maps. In

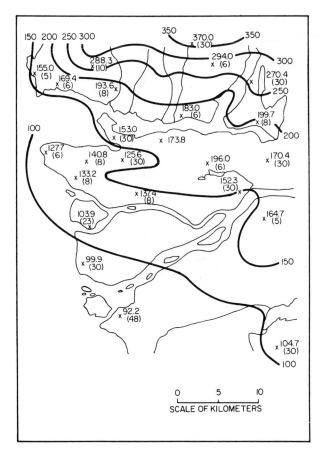

FIG. 22. Mean annual precipitation in centimeters for metropolitan Vancouver; the length of record at each station is in brackets (68).

fact, it is possible to achieve modest forecasting success by simply extrapolating the motions of low-pressure and high-pressure systems.

When weather charts are averaged over a month, much of the day-to-day detail disappears. The patterns then begin to show a close relation to latitude and to major geographic features such as the distribution of continents, mountains, and oceans. Even monthly mean cloudiness, obtained by superimposing daily satellite photographs upon each other, yields a strong "signal," despite large day-to-day variability. The wind field on this scale is often referred to as the *general circulation*.

The causes of weather anomalies are as yet poorly understood but it is clear that regional departures from average may affect conditions around the world. For example, a northward or westward shift of a subtropical anticyclone changes the storm tracks and the weather patterns

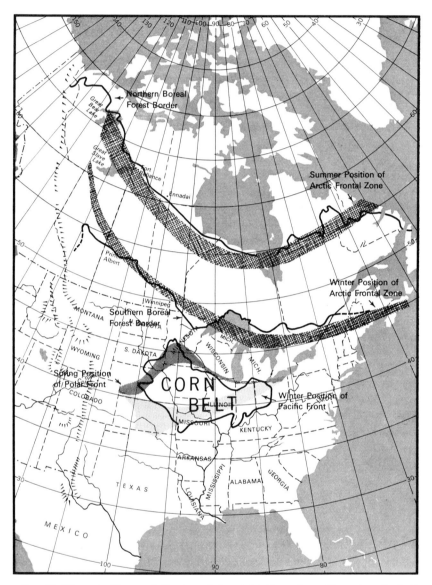

Fig. 23. The coincidence of the boreal forest biotic region with meteorologically defined climatic (air-mass) regions (70).

in temperate latitudes. Because living organisms are usually adapted to "average" environmental conditions, the biometeorologist is particularly interested in extremes, including the following:

(a) Short-term anomalies such as cold and heat waves, high air pollution potential, continental transport of pollen and spores, and distant travel of insects and birds.

(b) Seasonal anomalies such as drought, excessive wetness, and unusual ice-pack conditions.

(c) Long-term departures over decades or centuries resulting in major ecological changes.

Ecological climatology is the study of relations between climate and natural vegetation, as well as insect, bird, and animal distributions. In principle, soil and vegetation maps should be "mirrors of the normal climate of a region" (69, p. 953), and Hare (69) considers the search for precise relationships to be one of the most important challenges of modern climatology. Postglacial shifts in the major biotic regions have long been assumed to be associated with shifts in climatic patterns. As noted by Bryson (70), however, when climatic zones are deduced "from the biotic regions, and not derived independently, one is suspicious of circular reasoning in suggesting a causal relation."

Bryson has delineated climatic regions from analysis of air masses rather than from mean values of individual meteorological elements or from biological data. One of his results is given in Fig. 23, which shows that the average summer and winter positions of the arctic front correspond to the northern and southern borders of the boreal forest. The hypothesis of a causal relation is plausible. To the north of the arctic front in summer, for example, precipitation is uncertain and temperatures are not high enough to support tree growth, suggesting that the climatological position of fronts may be a useful index of the integrated effect of a number of weather variables. If the hypothesis is accepted, two interesting questions arise:

(a) Will the position of the tree-line shift if the general circulation changes, and if so, what is the time constant of the response?

(b) Will the general circulation change if the tree-line advances or retreats?

This initial statement on macrometeorology will be amplified in later chapters, particularly Chapters 10–13.

3.5. NETWORK SPACING

The three-dimensional atmosphere can only be observed imperfectly, and selection of sampler locations is an important and difficult problem. In most experimental studies, time continuity at a given point is easier to achieve than space continuity at a given time. Aliasing may occur in both cases; the abscissa in Fig. 6 can be labeled as distance rather than time.

For large international programs, Obukhov (71) remarks that network design should not be decided subjectively by majority vote of a committee. Instead, objective criteria must be specified, and preliminary feasibility studies are often required. This view is equally valid on the meso- and macroscales. Network design criteria depend upon the following factors:

(a) The purpose of the investigation.
(b) The nature of the underlying surface.
(c) The spatial variations of the sources of stress, e.g., the locations of chimneys in a city.
(d) The spatial variation in the number of receptors subjected to the stress.
(e) Practical limitations such as accessibility, availability of electric power, and protection from vandalism.

The purpose of the investigation is, of course, the primary consideration, determining the relevant scale of motion. Too close a spacing may introduce noise, and it is often desirable to use a smoothing technique. The meteorologist does this subjectively when he draws isobars on a synoptic weather chart. Too wide a spacing, on the other hand, may result in a partial loss of signal and aliasing. In many instances, a trial-and-error method is used, combined with some physical insight. The initial network is gradually modified after preliminary analysis shows that a particular station is providing duplicate information or, on the other hand, that large spatial gradients are incompletely observed. In some regional studies, measurements from permanent reference stations are compared with data from spatial traverses or from temporary stations, during steady-state macroscale weather conditions. A first estimate of frost climatology for a large rural area may be obtained this way.

If a substantial amount of money is being spent in establishing a network, it is desirable to increase the budget slightly in order to include a few additional stations or kinds of measurement. As the investigation proceeds, it is sometimes found that the original objective requires modification, that some secondary or unforeseen question becomes more interesting than the primary one, or that the data prove to be of interest to scientists in other parts of the world or perhaps in other disciplines. For example, a dense rain-gauge network established to test cloud-seeding theories may be useful also in hydrological or ecological studies.

The network may be required for one of three reasons:

(a) To obtain areal mean values, e.g., of watershed total precipitation.

(b) To interpolate between observing points for subsequent correlation with biological variables, or for provision of information on local climate to the engineer or to the farmer.

(c) To predict the behavior of environmental variables, using empirical methods or physical models that require knowledge of initial conditions.

In the last case, Obukhov (71) emphasizes that predictability is closely linked to the time and space scales being observed. For instance, if the distance between measuring points were only 1 mm and data were obtained every fraction of a second, well-defined spatial patterns would emerge, permitting accurate forecasts to be made for time periods of the order of a second or so. In this hypothetical case, the noise has become a signal, although there is still a random component, of scale smaller than the distance between observing points. Obukhov suggests that an indicator of the coherence of a particular meteorological field in space and time is the magnitude of the correlation coefficient obtained from measurements at different separation distances and for various averaging times. Frequently, the micro- and mesoscale correlations may vary with the macroscale patterns. Nevertheless, the coefficients and their variabilities provide a logical basis for network design. Taking short-range synoptic forecasts as an example, Obukhov reasons that the time and space scales, T and L, must obey the strong inequalities,

$$(3.2) \qquad\qquad T_G \ll T \ll 24 \text{ hr}$$
$$(3.3) \qquad\qquad H \ll L \ll L_0$$

where T_G is the characteristic period of a gravity wave and H and L_0 are characteristic vertical and horizontal dimensions of weather systems whose motions are to be forecast. Assuming that T and L are geometric means in Eqs. (3.2) and (3.3), e.g., $L = (HL_0)^{1/2}$, Obukhov obtains the values of 1 hr and 150 km for T and L, respectively, i.e., observations are required at hourly intervals from stations about 150 km apart.

The nature of the underlying surface is a factor in network design, particularly on the mesoscale. There is a natural tendency to take observations at equidistant grid points, but Landsberg (72) recommends that stations should be in lines at right angles to the main physical, biological, and cultural features (coasts, ridges, tree lines, built-up areas, etc.).

Spatial variations in the number of receptors may modify the basic network design. Vegetation species are often "clumped" rather than distributed randomly; human population is concentrated in cities and towns. The United States Atomic Energy Commission (73) suggests as a guide for establishing a monitoring program around a nuclear reactor that there be one air sampling station for each community of over 100 people within 8 km (5 miles) of the plant and an additional station for each town of over 1000 people within 16 km (10 miles).

A practical study of network spacing in Nashville, Tennessee has been reported in a series of papers (74–76). The final results are summarized in Table VIII, which lists the number of stations required to estimate spatial mean values for a number of pollutants and sampling times. Only a few observing sites are required to obtain annual values but a larger and quite variable number is necessary in order to estimate monthly means. Two-hourly spatial means of soiling index and SO_2 require 60 and 38 stations, respectively. Concentrations of SO_2 in summer are so low that only a very small network is required.

TABLE VIII

NUMBER OF SAMPLING STATIONS REQUIRED FOR MEASURING 2-HR, DAILY, MONTHLY, SEASONAL, AND ANNUAL MEAN CONCENTRATIONS OF CERTAIN ATMOSPHERIC POLLUTANTS IN NASHVILLE, TENNESSEE WITH ±20% ACCURACY AT 95% CONFIDENCE LEVEL[a]

Time period	Sampling stations required per 25.6 km² (10 mi²)				
	Sulfation	Dustfall	Suspended particulate matter	Soiling index	Sulfur dioxide
2-Hrly				60	38
Daily			9	60	38
Monthly					
Sept.	14	27	5	2	1
Oct.	17	25	3	4	10
Nov.	12	80	4	7	9
Dec.	6	41	2	14	8
Jan.	11	100	2	9	10
Feb.	8	16	2	9	12
Mar.	20	22	3	—	11
Apr.	7	18	3	8	5
May	14	20	2	10	1
June	7	11	2	6	1
July	9	20	2	7	1
Aug.	6	57	7	5	1
Season					
Fall	12	20	3	3	4
Winter	7	21	2	6	9
Spring	11	8	3	9	3
Summer	6	9	3	4	1
Annual	6	7	2	4	—

[a] Stalker *et al.* (76).

The Nashville study included a comparative health survey in different parts of the city. There was a need, therefore, to estimate not only the urban mean values but also the spatial gradients of pollution. Figure 24 compares the annual mean sulfation determined by networks of 32 and 119 stations. Although the patterns differ in minor details, the smaller network may be satisfactory for a health study.

A similar investigation in Sheffield, England (77) is in qualitative agreement with the Nashville results. Clifton et al. (77) suggest that only eight sampling stations are needed in Sheffield to estimate the average monthly values of smoke and SO_2 to within $\pm20\%$ of those obtained from a network of 28 stations. In order to obtain spatial gradients with reasonable accuracy,* however, the sampler spacing must be 0.8 km ($\frac{1}{2}$ mile) and 1.6 km for daily and monthly mean values, respectively. These estimates should not be extrapolated to cities with unusual spatial distributions of emissions, or with major orographic variations.

Finally, if a meteorological or air quality station must be moved because the land is expropriated or electric power is no longer available, there should be a period of overlap. This permits a comparison of the micro- or mesoscale influences at the two sites. In studies of long-term climatic or air pollution trends, however, every effort should be made to secure a 100-year commitment of land use for a reference station. In the case of a city, the permanent site for the air quality measurements should be within a large park away from the influences of automobile traffic, tall buildings, and very local chimneys. The same guideline does not apply for the climatological station. Landsberg (72) emphasizes that a large park creates a microclimate that is not representative of the urban environment. He suggests selecting an open site ranging in size from one city block to 1 km in diameter, completely surrounded by built-up areas.

Mesoscale frequencies of rare events such as tornadoes and hail are sometimes required by insurance companies. However, the observed average number of tornadoes per country, per state, or even per small unit area depends among other things upon the density of the observing network. The analysis is simplified by considering *tornado days* or *hail days,* thus eliminating the problem of not knowing whether two observers have seen the same or different tornadoes. Nevertheless, a voluntary network is rarely uniformly spaced and its reporting efficiency increases as the storm becomes more severe (78).

Carte (78) has analyzed data from an area in the Transvaal, South Africa, where there are about 800 voluntary hail observers. Figure 25

---

* The phrase "reasonable accuracy" will be defined in Chapter 6 in terms of the coefficient of geographic variation.

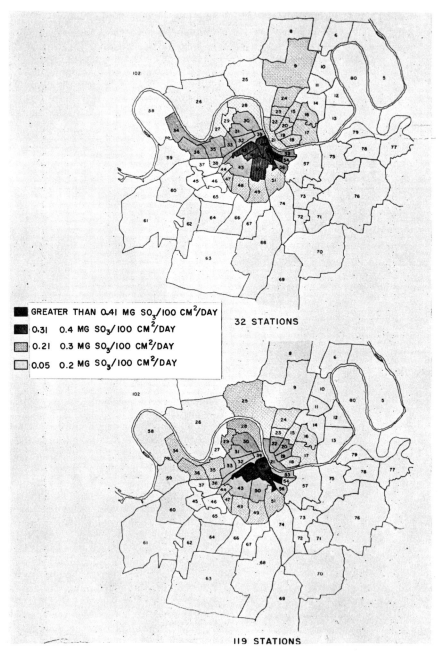

GREATER THAN 0.41 MG SO_3/100 CM^2/DAY

0.31   0.4 MG SO_3/100 CM^2/DAY

0.21   0.3 MG SO_3/100 CM^2/DAY

0.05   0.2 MG SO_3/100 CM^2/DAY

32 STATIONS

119 STATIONS

FIG. 24. Annual mean sulfation rate in Nashville, Tennessee, determined by networks of 32 and 119 stations (74).

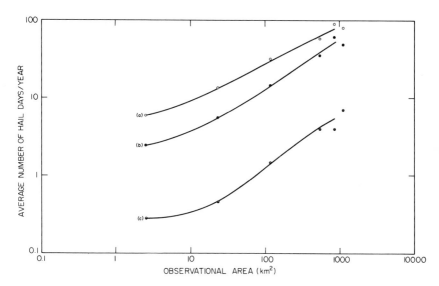

Fig. 25. Dependence of the average number of hail days per year on the size of the observational area; data are for the Transvaal, South Africa (78). The size of the hailstones were (a) any size, (b) greater than 1 cm, and (c) greater than 3 cm, respectively.

summarizes the results obtained during three seasons. In the case of the smallest observational area, 2.6 km^2 (1 mile^2), there were at least **5** observers in each square. The average number of hail days increases with increasing area but the relation is not linear. These results emphasize the difficulty of extrapolating areal averages from observations at a few points.

3.6. SOME INSTRUMENT SAMPLING PROBLEMS

Microscale spatial variability creates some special instrument problems. It is well known, for example, that most thermometers must be shielded and ventilated to avoid radiation errors. However, there is a danger within vegetation or in other confined spaces that forced ventilation may draw air from other heights; indicated gradients of temperature are then different from the true ones. More generally, Idrac (79) has used the term *finesse* to denote the ability of a sensor to measure without causing modification of the environment by its very presence. A medical thermometer does not have *finesse* if it provokes local irritation, causing the temperature to rise.

TABLE IX

THE RESULTS OF GLASS-BEAD DIFFUSION TRIALS AT SUFFIELD, ALBERTA[a,b]

Trial	Sampler height (m)	Percentage recovery from source to a downwind distance of 800 m
1	Surface	96
1	0.9	32
2	Surface	76
3	Surface	88
4	Surface	87

[a] Hage et al. (80)
[b] Median particle diameter was 108 μ; emission height was 15 m.

A flat-plate sampler, sometimes called a gravity slide, has been used traditionally to collect pollen or other small particles. The internal boundary layer (see Section 3.2) that develops over the plate, however, disturbs the air flow and the efficiency of catch. Hage et al. (80) have described experiments in which glass microspheres coated with fluorescent dye were emitted continuously from an elevated point source. A large number of samples were collected on clear acetate adhesive papers mounted horizontally at the ground and, in one case, at a height of 0.9 m; the sticky paper had an area of 33×16 cm^2 and was attached by masking tape to cardboard of area 40×20 cm^2. The recovery, defined as the ratio in percentage of the mass accounted for by the field samples to the total mass emitted, is given in Table IX; in Trial 1, the recovery at a height of 0.9 m was only about one-third of that at the ground. Examination of samples under a microscope showed a regular increase in deposit along the sticky paper in the direction of the wind, the effect being much more marked at a height of 0.9 m than at the surface. In addition, there was a variation across the sticky paper, with a maximum at the center and minima at the edges.

These results indicate that a flat-plate sampler should be positioned at ground level. Even so, the deposit is only qualitatively similar to that on a natural surface such as grass, which has a different roughness and efficiency of catch.

Another type of problem is illustrated in Fig. 26 (81). An aspirated sampler for particulate matter does not measure true concentration or dosage unless the air enters *isokinetically*, i.e., without acceleration or deceleration. This is a consideration not only for measurements in the atmosphere but also for those in dusty flue gases. Harrington et al. (81) have tested a number of pollen samplers, including an aspirated sampler with millipore filter, and sticky slides, pipes, and wires. A comparison is given in Table X.

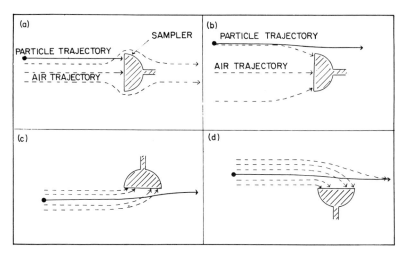

Fig. 26. Schematic diagram showing the effect of particle inertia on the collection efficiency of aperture-type samplers in a moving air stream: (a) sampler facing into a wind of higher speed than the flow into the sampler, (b) sampler facing into a wind of lower speed than the flow of the sampler, (c) and (d) sampler facing normal to the wind (81).

TABLE X

POLLEN CATCHES BY VARIOUS SAMPLERS IN A SLIGHTLY
TURBULENT WIND-TUNNEL FLOW[a,b]

Sampler	Facing	Orientation of long axis	Pollen count	Computed concentration $(g/m^3 \times 10^{-2})$
Slide	Downward	Normal to wind	0	—
	Downward	Parallel to wind	2	—
	Upward	Normal to wind	8	14
	Upward	Parallel to wind	744	1265
	Into the wind	Vertical	3850	58
Aspirated sampler with	Left wall		3653	307
Millipore filter	Right wall		9442	794
	Upward		5859	493
	Downward		4066	342
	Into the wind		73,500	6180
	Downwind		23,100	1942
Pipe (17.4-mm diameter)		Vertical	5470	92
Wire (1.0-mm diameter)		Vertical	3310	971

[a] Harrington *et al.* (81).

[b] Duration of test was 10 min, wind speed was 233 cm/sec, and the total volume sampled by the 6 millipore filters was 1.55 m³. The slides, pipe, and wire were covered with cellulose tape coated with dilute rubber cement.

In order to calculate concentration, the horizontal microscope slides were assumed to sample a volume equal to the product of the fall speed of the pollen (1.6 cm/sec), the area of the slide, and the time. The vertical slide, pipe, and wire were assumed to sample a volume equal to the product of the area of the sampling surface, the wind speed, and the time. The six millipore filters were all connected to the same aspirating system and each was assumed to sample an equal share of the total volume of air drawn through the unit.

Table X confirms that flat plates are inefficient collectors. Among other things, their catch is dependent upon the angle of attack of the wind. These results agree with those of Hage *et al.* (80). The data for the milli-pore-filter tests also show the effect of sampler orientation with respect to wind direction. The filter facing into the wind collected far more pollen than any other. In the case of a cylinder, the results in Table X suggest that efficiency improved with decreasing cylinder diameter. Microscopic examination showed that particles impinged directly on the windward side, the efficiency increasing with increasing particle size, and/or wind speed.

The wind direction is rarely steady outdoors, and particles are often deposited irregularly. Harrington *et al.* (81) have therefore developed the *rotobar sampler,* in which a bar coated with dilute rubber cement, 0.6 cm in diameter and 3.8 cm in length, is rotated at a constant speed. Particles impinge on the forward surface of the bar. This instrument was compared with others by Harrington *et al.* (81) at a height of 3.7 m above a weedy field during the normal ragweed season. The rotobar yielded pollen counts that were significantly higher than those obtained from a Millipore filter facing downward. A commercially available model of the rotobar, called the *Rotorod,* has been described by Leighton *et al.* (82).

Dingle (83) has discussed the 1946 recommendations of the Committee on the Standardization of Pollen Counting Techniques of the American Academy of Allergy. That Committee recommended that between two stainless steel disks 22.5 cm in diameter, positioned horizontally with one disk 7.5 cm above the other, a coated gravity slide should be placed 2.5 cm above the lower disk; the entire device should be mounted near the center of a large flat roof. Dingle (83) emphasizes that such a sensor can only be expected to yield qualitative results.

A final example of a microscale sampling problem concerns the collection of rain in gauges. Splashing and evaporation are possible sources of measurement error but the major concern is with the effect of wind and turbulence, which prevent droplets from entering a gauge isokinetically. An interesting exchange of letters to the editor of *Weather* appeared in 1958–1959 (Vol. 13, pp. 33–34, 210–211; Vol. 14, pp. 187–188,

367–368). As early as the nineteenth century it was known that the catch decreases with increasing gauge height and increasing wind speed. However, quantitative estimates of the error are difficult to obtain. Although Stanhill (84) found rather large differences (as much as 35% in strong winds) between values from gauges at heights of 30 cm (1 ft) and 2 cm (0.8 inch), Glasspoole (85) suggests that the site was "probably overexposed."

Provided that the gauge is at a distance of not less than twice, and preferably four times, the height of obstructions such as buildings and trees, the site should be in as sheltered a location as possible (86, p. 264). Over open prairie or at coastal locations, artificial shielding is recommended.

The spatial gradients of precipitation are often of interest. When there are significant differences in exposure, however, it is difficult to decide whether the measured gradient is valid. The determination of precipitation over lakes, using observations from shore stations and small islands, is uncertain, as is the estimation of rainfall around buildings. Lacy (87) measured the catch in 10 gauges positioned around a house. The results over 8 months, expressed as a percentage of rainfall at a "standard" exposure, showed values as low as 22%. Lacy concluded that only part of the deficiency was due to gauge exposure, the remainder being a real building shadow effect, but he was not able to be more precise.

4/THE DESIGN OF BIOMETEOROLOGICAL EXPERIMENTS

4.1. SOME PRELIMINARY STEPS

An experiment is undertaken to test an hypothesis (reasoning from the general to the particular—the *deductive method*) or to seek a relation (reasoning from the particular to the general—the *inductive method*). In biometeorology, the connections between stress and response are frequently very complicated; thus, there is often difficulty in formulating a hypothesis or in producing much more than a random scatter of points when biological response is plotted against each environmental variable in turn on a graph. Crop yields, for example, depend on antecedent weather conditions over a span of several months. In such cases, the following guidelines may be helpful in planning a biometeorological experiment:

(a) List the environmental factors that may be important.

(b) Attempt to identify appropriate space and time averages for each variable, although this may often prove to be the central obstacle towards progress, requiring a separate investigation.

(c) Undertake a literature survey, including a search of journals in related disciplines and from other countries. Even when previous relevant studies have produced conflicting results, the methodologies may be of assistance in the design of new experiments.

(d) Prepare a data inventory. In an ecological or hydrological study of a watershed, for example, every effort should be made at the outset to catalogue the sources of data. Many of the environmental factors are measured regularly by national weather services, agricultural and forestry departments, and public health agencies.

(e) Seek the advice of scientists in other relevant disciplines. The plant physiologist and the micrometeorologist should cooperate when choosing outdoor sampling locations, selecting the atmospheric variables to be measured, and deciding on the types of sensors to be used. If a statistical analysis is proposed, a statistician should participate in preliminary planning; if he is consulted later, he may find sources of bias that make the data useless.

(f) Consider whether it is possible to examine each stress separately, using controlled indoor laboratories (see Section 4.5).

(g) Consider the possibility of manipulating the natural environment (see Section 4.6). For example, the atmospheric dispersion of ragweed pollen must be understood if an urban weed control program is to be effective, but multiple sources make it difficult to design a suitable tracer experiment. To avoid this complexity, ragweed seedlings may be grown in greenhouses and transplanted to field plots in advance of the normal season. This method has been used at the University of Michigan and the Brookhaven National Laboratories (88).

(h) Take advantage of rare events for special studies. An industrial strike provides an opportunity for examining the effect of reduced emissions on local air quality. Similarly, if a large stand of timber is destroyed by fire, forest regeneration may be observed in an outdoor natural setting. Even a solar eclipse is of interest. Pruitt *et al.* (89) have studied the evapotranspiration of a vegetative surface at Davis, California during an eclipse on a cloudless day. It might be thought that local shading with an artificial screen would be equally effective but the air flow is then not in spatial equilibrium with the new surface energy balance; an internal boundary layer forms at the upwind edge of the shaded area.

(i) If large quantities of data are to be generated during the experiment, prepare in advance for the problems of sensor compatibility, data translation, data retrieval, and quality control. The computer is a useful tool but it should not dominate the design of an experiment. Too often the methodology is influenced by the availability of punched-card historical records and by the capabilities of computers. In urban studies, for example, a long homogenous record from an airport weather station does not necessarily provide a solution to every problem. Similarly, it may be convenient but not at all relevant to use data from an open grassland location in physiological investigations of adjacent forests.

Most textbooks on statistics include a chapter on the "design of experiments." In that context, the phrase refers to the problem of ensuring that samples are drawn at random from the same population. The Latin square method is an example of statistical design. Randomization is a specialized topic that will not be considered in this chapter except for occasional reference to sources of bias in biometeorological data.

4.2. VISUAL OBSERVATIONS

Of the five senses, the ability to see is undoubtedly of most importance to a biometeorologist. Many clues about environmental processes are on display, awaiting discovery by perceptive observers. Scorer (90), for example, suggests that soaring birds and dragonflies reveal the structure of convection below a cumulus cloud.

The studies of Okita in the Japanese city of Asahikawa illustrate the ingenuity of an investigator. A built-up area is warmer than the surrounding countryside (the urban *heat island*) (10, p. 199). The air therefore drifts in towards the center of the city when regional winds are light, but the flow is irregular and difficult to trace. To overcome the measurement problem, Okita (91) took advantage of the fact that on foggy mornings when temperatures were below freezing, rime ice formed on the windward side of tree branches. He therefore used the orientation and thickness of the rime formation as an indicator of the integrated local overnight wind flow. Some typical results are given in Fig. 27, which shows convergence towards the center of the city. On this occasion, February 26, 1959, the minimum temperature at the Meteorological Observatory was −17.9°C, whereas values at two rural stations 7 and 9 km distant were −20.4 and −20.0°C, respectively. In a subsequent investigation (92), Okita used the angle of drift of chimney smoke as a tracer of urban flow. A network of sensitive recording wind vanes would have been costly and would have yielded little additional information.

Vegetation patterns are useful in deducing climate, particularly on the mesoscale, although the absence of a species does not necessarily imply that environmental conditions are unfavorable: there may be competition from other species. Nevertheless, an ecologist can sometimes make a shrewd estimate of spatial climatic variations from an examination of the soils and natural vegetation. In an unpublished study of a number of Rocky Mountain Valleys differing in orientation, size, and slope, H. Cameron (personal communication) found that soil development varied throughout each valley in a far from random way. The patterns permitted him to envisage both the distant and the more recent heat and moisture budgets to which the parent material had been subjected. These inferences in turn could be explained by a consideration of physical controls such as aspect, slope, latitude, and altitude. To supplement the clues

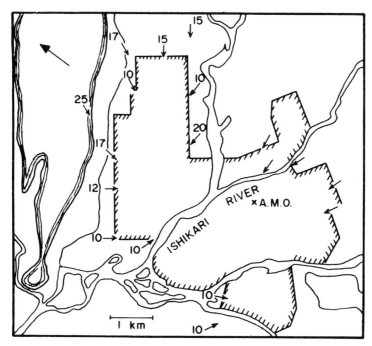

Fɪɢ. 27. Overnight wind flow at Asahikawa City, Japan, February 26, 1959 as obtained from rime ice on trees. The thickness of the rime ice in millimeters is given beside each arrow (91).

provided by soil types, vegetation distributions were useful indices. Long-term climatic controls were deduced from observations of the distribution patterns of soft woods and hard woods; short-term influences were inferred from a study of the predominating ground-cover species. The frequency of critical environmental events could be estimated qualitatively from an examination of growth rates, physical deformations, and die-back (frost kill and desiccation, for example). To an observant scientist, both current and historical environmental conditions are revealed qualitatively by observing the current state of nature.

Jenik and Hall (93) have evaluated the effect of the harmattan wind on vegetation in the vicinity of Djebobo Mountain (elevation of 876 m) in Ghana. In that region, a humid southwest monsoon blowing from April to August is replaced by a dry northeast harmattan from December to February. Jenik and Hall found isolated islands of savanna woodland growing on the leeward side (in relation to the direction of the harmattan) of each saddle in the ridges; the soils in these areas were comparatively deep. Windward slopes, on the other hand, were characterized by soil erosion, deficiency of soil moisture, and patchy stunted trees and shrubs. The authors suggest that the protection against desiccation afforded by the ridges plays a major ecological role in this region.

Wind-deformed trees are indicators of prevailing wind direction. Sekiguti (94) examined 230 persimmon trees distributed over 25 km^2 in Akaho Fan, Nagano Prefecture, Japan. He found that the local prevailing wind direction during the spring growing season could be determined from the permanent tilt of the trees with an accuracy of about $\pm 15°$. Sekiguti used this technique to describe the mesoscale prevailing wind patterns, and he was able to demonstrate the influence of valleys and escarpments in deflecting the regional air flow.

The angle of tilt of a tree has been suggested as an index of wind speed, but Thomas (95) emphasizes that other factors such as salt-induced deformations along a seacoast and soil depth variations are important. In order to evaluate the shelter-belt requirements of open agricultural areas in hilly country, Thomas (96) has mounted un-hemmed flags on metal rods at a height of 1.2 m. The extent of flag tattering, measured every few days, is an index of the total "run of the wind," thus permitting the mesoscale patterns to be mapped at little cost. Rutter (97) has found that the rate of tattering is a linear function of the run of the wind, except when the flags are wet or the ribbons are torn. He recommends that the flags be trimmed weekly.

Shibano (98) has studied mesoscale flows in the southern Boso Peninsula of Japan using not only tree-bending but also many natural and man-made indicators, such as the orientation of rice paddy furrows and of artificial shelter belts. Even regional variations in the structure of houses provided clues. In the Shirahama district, for example, concrete walls face south as a protection from typhoons; toward the coastline, the houses become more solidly built and hardly any straw roofs are to be seen.

Fog and smoke are tracers of local winds. Myers (99) has observed the fog patterns in a valley in Pennsylvania. His photographs suggest that trees and undergrowth inhibit the drift of fog from swamps and marshes and he suggests that shelter belts be used in certain instances to reduce the frequency of fog over highways and airports. Smoke from chimneys often yields a flow visualization of downwash while ignited oily rags and smoke bombs have revealed the principal features of valley winds. It is important, however, not to generalize from a few photographs because of the wind variability from day to day, and even from hour to hour.

Because oxidants cause rubber to crack, an inexpensive oxidant sampling network consists simply of rubber strips exposed outdoors. The samples are examined under a microscope at weekly or monthly intervals. Although the results are qualitative, they do indicate spatial variations of this kind of pollution.

Vegetation too may be used as a natural indicator of air quality, provided that the inherent uncertainties are recognized (100). Leaf damage is the integrated result of many environmental stresses including drought, wind, frost, insects, pesticides, plant diseases, and old age. In addition, only some species are sensitive to pollution, and they must be at a susceptible stage of growth for injury to be caused by a single fumigation. Nevertheless, field surveys by a plant physiologist are valuable because vegetation damage is frequently the first symptom of a local air pollution problem; some species show visible effects when contaminant concentrations are still too low for chemical detection. As hardier varieties begin to show injury, the trained investigator can judge the street-by-street extent of the problem. McKee and Bieberdorf (101) suggest that "trees work day and night as continuous sampling devices, and require no electric power and no maintenance."

In some community pollution studies, sensitive plants are grown in uniform soils that are regularly fertilized and moistened. Species are chosen which yield symptoms characteristic of specific pollutants, and in some cases, a laboratory bioassay is undertaken to determine the sulfate or fluoride content of a random sampling of leaves.

Phenology is the study of the timing of recurring natural phenomena such as the flowering of a tree or the freezing of a river. The Royal Meteorological Society, for example, published a Phenological Report annually for many years, which listed the flowering dates of a number of species in various parts of the British Isles. Jeffree (102) analyzed the records for the years 1891–1948 and found a marked latitudinal effect in all years: the mean flowering date, averaged over 11 species and all years, was 21 days later in north than in south Britain, a distance of 650 km. To examine year-to-year variations, he calculated the 58-year mean flowering dates at each location, and averaged the extreme deviations from these means, yielding a value of 17–18 days for both extreme earliness and extreme lateness. Finally, to test the relevance of the flowering event to other phases of plant development, Jeffree examined the record of the date of leafing of horse chestnut for the years 1929–1948. Only in 1938, 1941, and 1947 were there differences of more than 10 days (-10.3, $+15.0$, $+11.0$ days, respectively) from the dates inferred from adjacent flowering events. Jeffree therefore believes that the date of flowering is a useful phenological index.

Some indication of climatic trends may be obtained from phenological records. Arakawa (103) has published the freezing dates of Lake Suwa in Japan since the winter of 1443–1444, and the timing of full bloom of cherry trees at Kyoto, Japan since the eleventh century (104).

The phenological approach has been criticized for the following reasons:

(a) The particular date of flowering or of ice formation on a lake is difficult to specify. In some instances there may be a variation of 5 to 7 days in the estimates made by different observers.

(b) Phenology integrates the effect of many environmental variables, and the influence of an individual factor cannot be isolated.

(c) There may sometimes be genetic adaptations of vegetation with hardier varieties growing in less favorable climates.

(d) Although phenological records for a large area are usually compared with macroscale weather observations, the local microclimate controls the time of flowering.

Phenology is most useful in studying variations over a small irregular area. Geiger (105, pp. 386, 430, 439) has given a number of examples of this. On the microscale, blossoming on individual branches of a single pine usually begins on the south side of the tree.

On a larger scale, Jackson (106) has studied a forested area of 0.7 km² (180 acres), which includes three deep gorges. Sixteen microclimatic stations were each located at heights of 0.5 m near the sites where flowering dates were being observed. In the spring of 1963, there were large variations in both microclimate and flowering dates (full flowering was defined as the time when 90% of the individuals of a species had flowered within a 3-m radius of the station). For example, two stations 45 m apart on south-facing and north-facing slopes showed a difference of 6 days in the average date of flowering of 9 species. These variations seemed to reflect mainly the influence of cumulative degree-day differences. Jackson believes that "phenological research could be expedited by making observations in diverse microclimates during a few seasons rather than acquiring long-term phenological records."

Phenology can be used to demonstrate the existence of the urban heat island. Autumn frosts arrive first in the suburbs, and an observant layman can follow the weekly advance of frost damage into the built-up area. Similarly, in spring, the times of blossoming of fruit trees vary across an urban area. Figure 28 is a map of Cologne prepared by Miss Kalb (107) showing isopleths of the percentage of open apple blossoms on May 7–8, 1956: the urban influence is evident.

In this brief survey of phenology, only a few recent illustrative studies have been described. The bibliography is so extensive that many intriguing examples have had to be omitted.

Visual observations are also of use in hydrology. Lake-level extremes

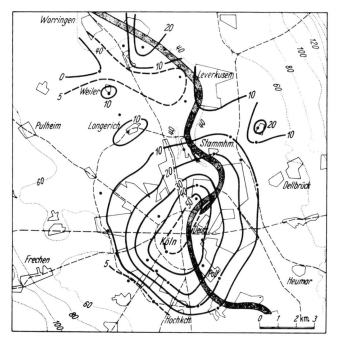

FIG. 28. Isopleths of the percentage of open apple blossoms on May 7–8, 1956 in Cologne (107).

can be inferred from an examination of shorelines, and in a study of maximum water discharges of the Tomin'ga River in the U.S.S.R., Levashov (108) noted the deformation of tree rings by ice in years of unusually large spring floods. In one case, he was able to infer that there was an ice run in the year 1768.

In a slightly different application, Saull (109) has noticed that blue-green algae flourish on water-covered bedrock only when there is sufficient sunlight. He suggests that the algae deposits in ancient rocks may perhaps be a climatic indicator.

In summary, visual observations are useful and inexpensive indices of bioclimate, assisting in the formulation of theories and in the selection of sites for more quantitative measurements. The environment often reveals its secrets if the investigator takes the time to look.

4.3. QUESTIONNAIRES

The questionnaire is useful in some human biometeorological investigations, particularly when there is difficulty in measuring quantitatively either the stress (e.g., odors) or the response (e.g., eye watering). The simplest way of assessing a person's discomfort is to ask him how he feels; his response integrates the effect of all the environmental stimuli. Questionnaires also can be used in the determination of the layman's interpretation of terms used in weather forecasts: what shades of meaning are conveyed to the public by the expressions *partly cloudy, cloudy,* and *overcast?* In a survey of students at the Pennsylvania State University (110), several anomalies were found. On a foggy day, for example, the majority selected the term "cooler," even on occasions when the temperature was actually slightly higher than on the previous day.

A questionnaire may range from a modest survey undertaken by a teacher among his students to a comprehensive sampling of a large population requiring a number of fulltime field workers. In all cases, it is most important to seek the advice of a statistician and a psychologist at the outset, not after the data have been collected; there are very subtle pitfalls to be avoided in designing a questionnaire and in choosing a population sample.

The individual questions should be as direct as possible, requiring only a single check mark or a yes–no answer. When volunteers are asked to provide daily information over a period of several months, the questionnaire must be short; otherwise, less and less attention will be given to accurate reporting as the weeks go by. In their analysis of a 3-month health study, Phair *et al.* (111) noted that 40% of the patients reported a progressive decrease in the day-to-day variability of their symptoms.

When only a single form is to be completed, the questionnaire may be longer. In this case, it is desirable to include control checks by either asking the same question twice, perhaps with slight rewording, or by asking two questions (not in succession), the answer to one being dependent in some way on the answer to the other, e.g., queries relating to occupation and income. Care should also be taken to avoid questions involving prestige or those which depend in some way on ethnic background. For example, respondents may tend to conceal their true cigarette smoking habits, if the questions are not carefully phrased. As another illustration, Anderson *et al.* (112) found that women between 45 and 54 years of age tended to state an age in the range of 35 to 44 to a census enumerator but were more likely to give their true age in a medical setting.

Broad questions should be asked before specific ones, and it is desirable to test the respondent's conception of general terms such as "weather modification" or "air pollution," giving him a choice of phrases. Care should be taken to avoid leading questions such as, "Are you in favor of reducing air pollution?"

Telephone interviews are sometimes used; however, in the St. Louis,

Missouri area in 1960, only 83% of the households had telephones (113), thus reducing the proportion of low-income families and nonfamily individuals in the sample. Face-to-face interviews, on the other hand, are sensitive to subjective variation, caused particularly by the tone of voice or facial expression of the interviewer. A printed questionnaire is often preferable.

The randomization of a population sample presents many difficulties. Anderson and Ferris (114) have offered a critique on community studies designed to probe the health effects of air pollution. Among other things, they note that a 75–80% response to a survey is inadequate because the noncooperative respondents often have an above-average incidence of symptoms. In addition, people employed in occupations exposed to respiratory irritants do not necessarily have an above-average incidence of respiratory disease. Except in small communities where there is no alternative source of employment, those who are adversely affected may have changed occupations. For similar reasons, the effect of shift work on health is difficult to determine because of natural selection factors; those engaged in night work may have less-than-average difficulty in readjustment of sleep patterns.

A single questionnaire may be used to determine spatial variations in stress and/or response. The results of one such investigation are given in

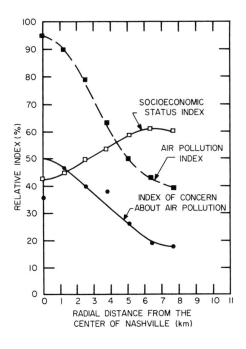

FIG. 29. Indices of concern about air pollution, of socio-economic status, and of air quality as functions of radial distance from the center of Nashville, Tennessee (115).

Fig. 29, which shows indices of socioeconomic status, air quality, and concern about air pollution as functions of radial distance from the center of Nashville, Tennessee (115). The trends are as might be expected. Another conclusion of the same study is that public concern seems to be influenced more by the frequency of extremes rather than by high monthly, seasonal, or annual average levels of pollution.

Paulus and Smith (116) designed a questionnaire to probe temporal variations among students with a recent history of allergic bronchial asthma at the University of Minnesota. The respondents kept daily asthma diaries, which were used subsequently to seek relationships with meteorological and air quality factors. Since there was a suspicion that dust from grain mills was a contributing factor, each student was given a skin test to determine whether he was sensitive to that kind of dust. One of the results is given in Table XI, in which the 1964–1965 academic

TABLE XI

AVERAGE NUMBER OF ASTHMA ATTACKS PER DAY PER STUDENT AMONG A
GROUP OF VOLUNTEERS AT THE UNIVERSITY OF MINNESOTA:
ALL HAD A RECENT HISTORY OF BRONCHIAL ASTHMA[a]

Group	Favorable 68 days[b]	Less favorable 157 days	Number of students
Grain-positive students[c]	0.052	0.032	51
Grain-negative students[d]	0.034	0.028	76

[a] Smith et al. (115).
[b] Days when the smoke index was high and the wind was blowing from the direction of the grain mills.
[c] Statistically significant difference.
[d] No significance.

year has been divided into subgroups of 68 days (when the wind was blowing from the direction of the grain mills, and the smoke index was relatively high) and of 157 other days. For the grain-sensitive students, there was a significant increase in the incidence of asthma while the remainder showed little change.

4.4. OTHER MEDICAL STATISTICS

Hospital admittance records have been used for studies of meteorotropisms. In some countries, however, there is often a delay of several days in obtaining a hospital bed; thus, the date on the admittance record may not correspond to that of the meteorotropism. Another difficulty is that the initial diagnosis may be incomplete. McCarroll (117) has emphasized that during spells of poor air quality, visits to hospital wards reveal no distinctive air pollution symptoms; instead, there is an increase in the number of respiratory and cardiac cases. Bridger and Helfland (118) have discussed a related problem, the classification of cause of death. Although a summer heat wave may increase the death rate, individual certificates may not list intense heat as the primary or even the secondary cause.

Bouchtoueva (119) has shown considerable insight in her discussion of methods for seeking relations between health statistics and air quality levels. An objective of such investigations is the determination of maximum permissible dosages over long periods of time. Indoor limits have been established for many contaminants but there is little information on the physiological effects of the more moderate lifetime dosages from mixtures of pollutants experienced outdoors in towns. The latter question has been studied in four different ways:

(a) *Spatial comparisons of total morbidity.* In this approach, morbidity rates in different cities or in different parts of the same city are related to pollution levels in the respective areas. However, the populations differ in many respects other than the quality of the air they breathe. The value of the socioeconomic status index plotted in Fig. 29, for example, varies across the city of Nashville. Not only is the air more polluted in the central core, but also the inhabitants on the average eat less nourishing food and live in overcrowded dwellings. There is no way of proving, therefore, that pollution is a cause of increased morbidity, even if in fact this is so.

(b) *Spatial comparisons of the health of children.* Children are likely to form a more homogeneous sample than the general population, and may be more responsive to some kinds of pollution stress. Again however, Bouchtoueva questions the validity of spatial comparisons when the parents belong to different social, economic, and occupational classes. Useful studies have been undertaken when a particular contaminant occurs alone and at sufficiently high concentrations to produce clinical symptoms. Examples include alterations in the lungs of children living near an excessive source of SO_2, and the mottling of dental enamel in an area polluted by fluorides. However, the method usually fails in areas of light to moderate pollution, and therefore cannot be used to establish maximum permissible dosages.

The detection of compensating reactions is a more promising method, i.e., the study of protective inhibitions that commence in the organism at the very lowest level of pollution stress. Examples include changes in the olfactory thresholds and a decrease in the excitability of the auto-nomic nervous systems of children.

(c) *For a single population, studies of day-to-day morbidity and mortality in relation to day-to-day air quality.* An example of this method was given in Fig. 3, Chapter 1. Because only one population is under investigation, the criticism leveled at approaches (a) and (b) above does not apply. However, Bouchtoueva suggests that the study of meteorotropisms does not reveal the long-term effects of low dosages.

(d) *Space and time studies of allergic reactions to pollution.* Dr. Bouchtoueva believes that health studies of people subject to skin allergies and asthma are most likely to be successful in establishing threshold concentrations of pollution. Evidence is accumulating that allergic reactions increase in a reproducible fashion as air quality deteriorates. The results of one such study were given in Table XI, and more investigations are recommended.

4.5. CONTROLLED LABORATORY EXPERIMENTS

In cases of multifactorial stress, the investigator often tries to isolate the effect of each variable in turn. This is not usually possible in the natural outdoor environment. On mountain tops, for example, atmospheric pressure is lower than in valleys but temperatures are also lower and winds are stronger. A controlled laboratory experiment is therefore an attractive alternative for examining the specific effects of reduced air pressure on physiological indices.

In other cases, the stress is so slight that a season or many years may be required for a measurable response. The *acceleration* principle is then used sometimes; one of the environmental stresses is intensified to shorten the experiment, as in corrosion studies. Alternatively, a more sensitive receptor may be substituted, or one which has a shorter life span, e.g., a guinea pig in place of a human being.

The central question to be answered in each experiment is the relevance of controlled laboratory studies to natural conditions. For example, the spectrum of solar radiation is never completely simulated in a growth chamber; this may not always be a deficiency but in any event, the effect of artificial daylight on the growth of seedlings must be determined before introducing other stresses.

Another factor often overlooked is that living organisms have become adapted to environmental variability in many cases. Akerman (120) suggests that students work better in classrooms under programmed fluctuations of indoor climate than in a constant environment. Similarly, Evans (54) described some experiments in which tomatoes were grown at a temperature of 22.5°C. In one treatment, air temperatures were kept steady while in other plots, controlled cyclic fluctuations were imposed with periods of about 2 min and amplitudes of 0.5, 1.5, and 2.5°C, respectively. There was a significant increase in both leaf area and dry weight with increasing temperature amplitude. Although this particular experiment may seem to suggest criticism of controlled laboratory studies, it in fact demonstrates the potential value of such work; how could the effect of temperature fluctuations be determined outdoors?

Andersen (121) believes that in the evaluation of human response to indoor climate at least, experiments in classrooms and offices are superior to those performed under laboratory conditions. He prefers to work with children, comparing groups in older schools with those in newer ones. In order to interpret the physiological responses, relevant measurements of the classroom physical environment must be included, as Andersen emphasizes.

The wind tunnel is used primarily to model physical rather than biological processes. Applications have included simulation of the wind flow over Gibralter, ventilation of the British House of Commons, the downwash of smoke behind buildings, snowdrift patterns, and the wind loading on tall, flexible structures. Some tunnels such as those at New York and Colorado State Universities include heating and refrigeration units to simulate vertical temperature gradients. A view of the University of Western Ontario installation is shown in Fig. 30, where the effect of turbulence on a building is being studied. Water may be used in place of air, for example in the dishpan experiments that model the global general circulation, provided that the scaling criteria are understood. The physical basis for modeling is considered in Section 8.3.

Greenhouses have been used for many years for experimental studies of plant growth. Some control is possible, for such variables as air temperature, humidity, and photoperiod. However, there is a danger that the data will be misinterpreted because of an imperfect knowledge of all the relevant environmental variables. Although the possibility of frost is eliminated, and although the air may be purified to remove pollution, the gradients and fluxes within and above the plants do not always resemble those outdoors.

Within a heated unventilated greenhouse in Ottawa during cold winter weather, Desjardins and Robertson (122) found large horizontal and vertical temperature gradients, as much as 5°C during sunny weather. Their results could be explained partly by differences in solar radiation; at two positions at the same height in the greenhouse on a

FIG. 30. A view of the inside of the wind tunnel at the University of Western Ontario; the effect of turbulence on a building is being studied. (Photograph courtesy Professor A. Davenport.)

sunny March day, one position received 50 ly more than the other in the morning but 40 ly less in the afternoon.

The phrase "wind tunnel experiment" usually implies that the investigator is working with miniature replicates of real objects. However, Hunt

FIG. 31. Outside view of the Australian phytotron CERES (124).

et al. (123) have used a tunnel to study photosynthesis and evapotranspiration of stands of alfalfa and orchard grass under controlled wind and radiation regimes. The working section was $30 \times 30 \times 90$ cm. Twelve incandescent lamps were submerged in a clear Lucite tray of distilled water at the top of the tunnel; the water acted as an infrared filter but some difficulty was encountered with a "sky" temperature higher then outdoors. Nevertheless, vertical profiles of wind, CO_2, and water vapor showed the expected gradients. This experiment included controlled changes in the flow rate, permitting the measurement of small wind-induced changes in the rate of net photosynthesis, an investigation hardly possible outdoors.

The *phytotron* is a building containing a large number of controlled "climates" for the study of plant growth. Figure 31 is a photograph of the Australian phytotron, called CERES (Controlled Environment Research Laboratory). Morse and Evans (124) suggest that an installation of this size has four advantages over smaller chambers:

(a) Experiments can be undertaken simultaneously over a wide range of atmospheric conditions, permitting rapid evaluation of a hypothesis. If only a single chamber were available, different batches of seedlings would be required for successive plantings, introducing uncertainty.

(b) Overhead maintenance and monitoring costs are reduced.

(c) More flexibility is possible, with less "down time."

(d) The phytotron facility attracts a relatively large number of scientists from different disciplines, frequently increasing research productivity.

The Australian phytotron has provision for three levels of control. First, there are chambers which use natural sunlight; the temperature may be adjusted to any value within the range 15–35°C with an accuracy of ±1.5°C under full solar load; the plants may be wheeled on trolleys to dark rooms to vary the number of hours of daylight. Second, there is intermediate control; these chambers are also naturally lit; however, the temperature range extends from 1 to 40°C, with an accuracy in any horizontal plane of 0.25°C. Third, there are artificially lit cabinets with very precise temperature and humidity controls and relatively uniform light intensities. The ventilation rate is about 4 times the rate that would be required if the plants were able to use all the CO_2 in the air; the flow in the cabinets is vertically downward at a rate of about 0.5 m/sec (100 ft/min), which is rather different from outdoor conditions. A feature of the installation is the ease with which environmental conditions can be changed.

Adams (125) has described an *air pollution phytotron* at Washington State University. He notes that if the flow were horizontal, the pollutant would pass over each plant sequentially, and concentrations would decrease downstream. To avoid the resulting interpretative uncertainties, the air motion is therefore from ceiling to floor, as in the Australian phytotron. The chamber is artificially illuminated and contains the usual temperature and humidity controls. When fluoride is the pollutant under investigation, the spatial gradients in the phytotron are determined very simply by hanging sets of lime-treated filter papers at various locations.

The University of Wisconsin *biotron* (126) is a building containing a large number of controlled climates for the study of plants, animals, and humans. The 48 laboratories include the following:

(a) A large-animal chamber with a temperature range from —25 to +45°C, high light intensity, and a wide range of humidities.

(b) A wind tunnel with a temperature range from —20 to +45°C.

(c) A desert-climate room with a capability for maintaining low relative humidities and high temperatures (up to 55°C); the barometric pressure can also be varied.

(d) A temperate-climate room in which barometric pressure may be varied between 920 and 1050 mb.

(e) Relative-humidity rooms in which high-order control is possible, for use in studies of leaf wetness and dew.

(f) A number of controlled animal-rooms.

(g) Plant nurseries.

(h) A cross-gradient room for studies of the effect of horizontal gradients of temperature and light on plants and animals.

The biotron has many automatic control features. Values of any of the environmental variables can be changed by very small amounts at time intervals of 10 min.

Although the usefulness of phytotrons and biotrons for interdisciplinary studies is evident, the work of the individual investigator should not be discounted. Laboring alone with perhaps only a few hundred dollars worth of "bits and pieces," his contribution may nevertheless be considerable. A rather simple laboratory device is the humidity-gradient cylinder, which is used for determining the preferred environmental habitat of insects. The cylinder is usually about 100 cm in length and 5 cm in diameter. Each 10-cm section contains a false porous floor, beneath which is a salt solution for maintaining the desired relative humidity. A temperature-gradient cylinder may also be constructed readily. Reichle (127) has described experiments in which 5 species of bog beetles were released at random positions along the cylinders, and their motions noted at 5-min intervals for half an hour. Only five beetles were permitted in a cylinder at one time, to avoid group behavior. All species selected a relative humidity of 95–100%, but there were significantly different responses to temperature; the lowest and highest preferred values were 19.5 ± 0.86 and $28.5 \pm 0.55°C$, respectively. Field studies of a bog in Illinois subsequently demonstrated population distributions in agreement with the laboratory findings.

Special precautions must be taken in laboratory investigations of living organisms. A stimulus above threshold may cause the organism to become "adapted" so that subsequent stresses have a reduced effect; this is particularly true in studies of sensory perception of odors and pain. Furthermore, care must be taken not to introduce associated stimuli or clues. Buchberg et al. (128) have described an experiment designed to determine the response of volunteers to eye irritants. It was necessary to eliminate the following: (a) odors, (b) temperature change between purified and chamber air, (c) change in flow rate when polluted air was introduced, (d) visual clues from other volunteers, and (e) patterns in the order or timing of stimuli which might allow the volunteer to anticipate.

A climatic chamber called a *climatron* has been used at the University of Pennsylvania to investigate rheumatoid arthritis (129). Patients who claimed to be weather sensitive were placed in the chamber in pairs for periods of at least 2 weeks. During the first 5–7 days, climatic factors were kept constant until the patients became fully adjusted to their environment. Without warning then, the pressure, temperature, humidity, and ionization of the air were changed singly and in combinations. With the exception of one patient, a variation in one of the elements had no effect. However, a combination of increased humidity and decreased pressure worsened the arthritic condition, as indicated in Table XII. In three other experiments, a low pressure and high relative humidity were maintained for 24 hr: following an initial increase in symptoms, the condition of the patients began to improve.

The *thermoheliodon* was developed at Princeton University for labora-

TABLE XII

EFFECT OF INCREASING HUMIDITY WITH FALLING BAROMETRIC PRESSURE
ON THE SEVERITY OF RHEUMATOID ARTHRITIS[a]

Patient	Total trials	Increased arthritis	Percentage
B.B.	7	4	57
F.F.	11	8	73
D.M.	6	4	67
E.L.	6	0	0
R.M.	5	3	60
D.N.	3	2	67
L.L.	3	3	100
E.A.	5	5	100
Total	46	29	63

[a] Hollander and Yeostros (129).

FIG. 32. The thermoheliodon at Princeton University for laboratory study of the heat loading on buildings (130).

tory study of the heat loading on buildings (130). Figure 32 is a photograph of the facility. Direct solar radiation is modeled by moving a 5000-W bulb along an adjustable track, thus simulating the effect of latitude and season. The acceleration principle is used, i.e., a 12-hr day is scaled down to 20 min, increasing the rate of experimentation. In this application, there is no need to model precisely the spectrum of solar radiation, particularly the photosynthetic wavelengths. Ventilation is provided to model convective heat transfer. Nevertheless, temperatures are kept well above room values to simulate long-wave radiative transfer to the cooler dome above. Diffuse solar radiation is modeled with four 300-W bulbs covered by grills.

4.6. OUTDOOR QUASI-CONTROLLED EXPERIMENTS

An attractive alternative to indoor laboratory experiments is *in situ* quasi-controlled field work. This may take the form of sudden or episodic stress imposed on a living organism in its natural environment when weather conditions are steady for an hour. As illustrations, a seedling may be exposed to frost in midsummer by deliberately chilling the air in a temporary insulated enclosure (131); a few branches of a tree may be enclosed in a tent and fumigated with a high concentration of SO_2 (132); a human volunteer may operate a treadmill in the desert during hot sunny weather to evaluate heat stress (133).

In studies of forest productivity, a plastic ventilated tent is often placed over a large branch, and CO_2 concentrations are monitored at both the intake and outlet of the air stream. However, the micrometeorological environment is seriously disturbed after only a very few minutes. Larcher (134) suggests that the temperature of a twig may become 10°C higher than it would be if the enclosure were absent.

Over longer times, the natural environment can be controlled somewhat by irrigating, shading, or mulching the soil. The results are difficult to interpret, however, because more than one of the environmental factors may be affected. Artificial shading of a plantation of trees changes not only the radiation balance but also the leaf temperature, transpiration rate, and soil moisture. As another example, when an animal population is deliberately introduced into an enclosed area in a study of productivity, a long chain of environmental reactions is generated: overgrazing worsens the microclimate which in turn diminishes the food supply. These interactions among the environmental factors do not always destroy the value of an experiment, provided that they are recognized and monitored. In some cases in fact, secondary responses of the ecosystem may be of primary interest.

Instead of attempting to modify the microclimate, the investigator can apply the method of *differential control*. The simplest example of this principle is *replication,* the exposure of a number of similar organisms to the same natural conditions. Although no control is exercised over the environment, an effort is made to subject all the samples to the same stress. One seedling may die but statistically significant statements can be made about the population from which the random samples are drawn.

Differential control also includes planting of replicated seedlings in adjacent fields, only one of which is fertilized or irrigated. However, the microclimatic variations that may occur are not always recognized. A small depression of only a meter or so in depth may be a frost pocket at night; in other instances, adjacent trees or buildings may change the character of the air flow and of the energy balance. Thus, two fields that look alike may in fact be quite different. Too often the published record of an investigation describes in detail the types of fertilizer and frequency of irrigation but fails to mention whether there was microclimatic uniformity.

Comparative experiments on seedlings are often replicated over 5–10 years in order to smooth away year-to-year weather variability. Alternatively, the productivities in individual years are related to climatic anomalies. In the latter case, the investigator should remember that plants have rather broad "comfort zones." Many of the environmental needs (such as water and sunlight) are often in great abundance. The survival and growth of the organism then depends on only a few limiting factors. These factors should be identified, and their stress thresholds determined, both separately and severally. A mean summer temperature that is 2° below normal may have little effect on the harvest but a departure of 6° may be disastrous.

Differential control also includes transplanting to a completely different continent or climatic zone (*provenance* tests). There are, however, usually too many uncontrolled variables to obtain a physical insight into the nature of the stresses, although superior strains may be found by trial and error for local use. In this connection, McFadden (135) has found that seeds may have been preconditioned in some way by their home environments: "there is no assurance that yields obtained will always indicate a true genetic potential for all varieties."

Emphasis thus far has been placed on the importance of seeking horizontal uniformity in experimental sites. A much more subtle method is to choose areas containing spatial gradients of the species and/or of environmental factors. For example, oscillations of ecological boundaries have been studied. Grime (136) points out that this is laborious, requiring monitoring over many seasons: furthermore, in the case of vegetation at least, the movements of frontiers may depend on rare events such as late spring frost, drought, or browsing by wild animals.

In the central part of a forest, there is little variation of vegetative growth from year to year. Trees in this region are called *complacent,* a term indicating the existence of environmental equilibrium year after year; the forest has created its own favorable microclimate. At the border, on the other hand, annual growth is dependent on climatic anomalies. Trees near the edge are termed *sensitive* and may be used as indices of climatic variation.

Clements (137) has suggested that environmental microscale gradients be exploited for physiological and ecological purposes. A frost pocket and a forest clearing both provide striking spatial contrasts in climate against a common background of mesoscale and macroscale weather conditions. Duffy and Fraser (138) have described a frost hollow at Centre Lake, Petawawa Forest Experiment Station in Ontario, Canada. The depression is a kettle 91 m wide and 9 m deep, containing a number of very old but stunted spruce and balsam fir. Here, between 21 June and 7 September 1956, frost occurred on 23 nights at the bottom of the depression but only on 5 occasions at the rim. Other examples have been discussed by

Geiger (105, pp. 393–403) and by Hawke (139). One of the more famous
hollows is the Gstettneralm doline in Austria; on one clear winter night,
the temperature was −28.8°C in the center of the depression but −1.8°C
at the rim, 150 m higher. Many of the features of alpine plant physiology
and ecology can be studied in such an environment.

Clements (140) has been "calibrating" a forest clearing as an outdoor
growth chamber. The soil moisture and radiation gradients across a
clearing are very sharp, permitting study of the differential growth rates
of seedlings in a natural environment. Figure 33 shows the actinograph
records on a sunny day in different parts of a clearing (61 × 15 m) at
Petawawa, Ontario. This approach merits wider consideration.

There are many other possibilities for outdoor quasi-controlled experiments. For example, Carson et al. (141) have described a Chicago fuel-
switch test. In late June 1968, many of the large industrial and all of

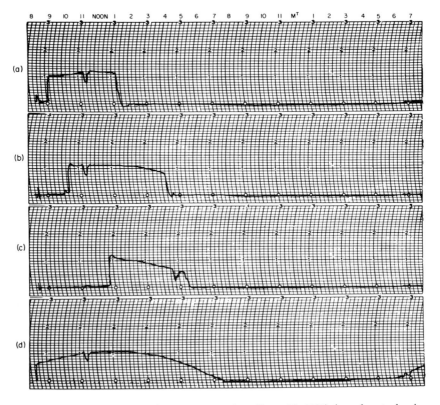

Fig. 33. Actinograph records on a sunny day (June 30, 1965) in a forest clearing
and in a nearby field at Petawawa, Ontario. Units of the vertical scale ly/min: (a)
west side; (b) center; (c) east side; (d) open field (140).

the utility sources of SO_2 increased their emissions by burning high-sulfur coal for a short period at the request of the Department of Air Pollution Control. This was a planned program designed to test urban diffusion models.

In studies of pollution from a single chimney, it is often preferable to release a specific tracer such as uranine dye rather than to utilize the pollutants emitted routinely; background contamination is thereby reduced, and the threshold value capable of detection may be lowered. Similarly, indium has been injected into clouds as a tracer of convective motions (142). Radioisotopes offer many possibilities for following pathways within an organism and for estimating transfer rates at the environmental interface (143); thus, for example, tree-root systems can be mapped without excavation.

4.7. STUDIES BASED ON CLIMATOLOGICAL AND OTHER ROUTINE OBSERVATIONS

Sometimes an investigator has no control over the siting of stations, the types of instruments, and the frequency of observations. Many studies of meteorotropisms, for example, are based on airport synoptic data, which represent incompletely the weather conditions in built-up areas.

These deficiencies should be considered as a challenge and should encourage biometeorologists to request that national and international organizations be more responsive to the data collection and dissemination needs of the life sciences. Some problems cannot be solved without 50–100 years of special observations. This is all the more reason for beginning a regular program as soon as possible, including adequate procedural documentation. National meteorological services maintain station registers, containing careful notes on changes in exposure, instrumentation, times of observations, and other details. This practice has not been followed generally in related disciplines. The original records or samples also should be preserved, wherever possible. In the case of air pollution, the high-volume sampler collects suspended particulate matter which is almost indestructible; the daily samples may become of irreplaceable value in future decades if new analytical methods are developed or if new trace substances assume importance in health studies. Similarly, biological "banks" have been proposed for the preservation of living species that are almost extinct.

When beginning a study based on routine observations, the investigator should attempt data *normalization*, the transformation to relative time or space scales in order to provide more generality. A trivial example is the conversion of a travel distance and a travel time to a velocity, i.e., normalization to unit time.

Normalization is frequently encountered in biometeorology (see, for example, the *rem* unit mentioned in Section 2.3 and the *clo* unit in Section 8.7). Of particular interest in the analysis of climatological data from various seasons and latitudes is the transformation from local to biologically oriented times. Examples are contained in the studies by Robertson (144) and Singer and Raynor (145).

Robertson (144) has suggested that because of the large latitudinal variations in photoperiod, the growth of wheat should be related not to calendar dates, but to phenological events: planting, emergence, jointing, heading, soft-dough stage, and ripening. The effects of moisture and temperature anomalies may then be related specifically to each stage of development of the crop (see Section 11.5).

Singer and Raynor (145) have proposed a solar time classification, to be used in the study of diurnal cycles, e.g., in ornithology and air pollution. For many living organisms, the important reference times are sunrise and sunset. These vary with season, with latitude, and, in a particular time zone, with longitude. Singer and Raynor have therefore divided the day into four periods: midnight to sunrise, sunrise to noon, noon to sunset, and sunset to midnight. If sunrise is at 7 AM and a biological event occurs at 9 AM, for example, the normalized time is +120 min in period two.

The merit of this classification is that micrometeorological conditions are relatively constant near noon and midnight but change rapidly at sunrise and sunset. The calendar of Singer and Raynor *normalizes* the diurnal cycle for time of year and latitude, permitting comparison of biological events.

5/TABLES, GRAPHS, AND CHARTS

5.1. INTRODUCTION

Tables, graphs, and charts assist in the search for biometeorological relationships among "noisy" and sometimes partially irrelevant data. Visual aids are helpful at all stages of an investigation, from the initial design to the final presentation of results. This is true both for the "armchair" study and for the experimental field investigation.

The most important sources of tabular information are the National Weather Services, which publish regular climatological summaries. The tables have been designed to meet the common-denominator needs of users in broad fields such as agriculture and aviation. The individual observations have usually been given many internal quality controls and are as free from error as is humanly possible. In the United States, for example, daily maximum temperatures are compared with values from adjacent stations in a cross check for inconsistencies. Nevertheless, the summaries are not always in the most useful form for particular biometeorological studies. For example, the class ranges in the published frequency distributions may be too wide, or a contingency table relating two specific variables may not have been included. In such cases, an inquiry to obtain data should be directed to the appropriate National Weather Service. The information may be available in unpublished form or may be obtained readily from punched-card archives.

Biological and air quality data are not so easy to find, and there are wide variations in observing, archiving, and publishing procedures. For example, records of hospital admissions, including initial diagnoses, are placed on punched cards in many parts of the world but the reporting and coding methods vary. The investigator must often undertake at least a partial quality control.

In the case of experimental programs, tables and graphs are helpful as quality controls of sensor signals and recording systems. It is often difficult to detect a calibration drift by visual inspection because of weather variability. Instrument checks are essential but in addition, the experimenter should tabulate or plot the data regularly as a further control. If analysis is not begun until after completion of a 6-month field study, for example, some of the data may prove to be worthless. This is a particular hazard when the observations are on magnetic or paper tape with no opportunities for visual inspection.

After a few data have been collected, preliminary tabular or graphical analysis often provides guidance, confirming the original experimental design or suggesting modification. This is particularly relevant if derived quantities such as an index of heat stress are to be used.

Tables and graphs assist in the search for relationships among variables. Most of the preliminary analysis will be relegated eventually to the wastebasket, although there is an understandable reluctance to discard calculations or summaries that may have taken many hours to complete. Nevertheless, if an investigator can proceed directly from experimental design to final result without swerving, he probably is not doing research.

Another valuable tool is *cartography*, which is the "art" of drawing isopleths for values of physical and biological variables in two- or three-dimensional space. The resulting chart is a form of data summarization.

Finally, visual aids assist in the meaningful presentation of results.

5.2. SINGLE-VARIABLE TABLES AND GRAPHS

Many biometeorological variables are not distributed normally, and the arithmetic mean is not the most probable value. For instance, the cumulative frequencies of air pollution measurements from an urban station are straight lines (approximately) when plotted on log-normal paper. Even when the Gaussian law is closely followed, the tips of the bell-shaped tails may be missing. This is not because of insufficient observations but because of physical constraints. Wind speed fluctuations over an hour or so, for example, are often distributed normally but extreme values are suppressed. Tables and graphs are therefore useful for identification of the nature of distributions and for determination of the most appropriate indicators of central tendency, of dispersion and of extremes, an important preliminary step in many biometeorological studies.

As a guide (146, p. 13), the number of classes in a table should be less than five times the logarithm of the number of observations, i.e., sets of 50, 500, and 1000 observations should not be subdivided into more than 8, 13, and 15 classes, respectively. The lowest and highest values may also be listed, but if the extremes vary widely from period to period, the 5% and the 95% values in the cumulative frequency distribution may be more meaningful.

Another single-variable tabulation is that of persistence, such as the frequency distribution of the length of frost-free periods, or of the number of consecutive hours that the wind direction is steady. In order to be meaningful, there must be no missing data. Furthermore, the tables require careful interpretation. In a listing of the number of consecutive days with measurable precipitation, for example, there may be instances of a few hours of rain overlapping two observing days. If the rain had begun an hour or so later, or ended an hour or so earlier, the series of rainy days would have ended.

Practical considerations are sometimes of overriding importance in the preparation of a duration table. For a large optical telescope, for example, the designer had specified that no astronomical measurements could be made when the winds exceeded 13.5 m/sec (30 mph), even if skies were clear, because of instrument vibration. Duration frequencies for these conditions at the proposed site were therefore requested from the meteorologists, but for hours of darkness only, with the added rather indefinite criterion that whenever the wind fluctuated about the critical speed for a few hours, the astronomers would probably not try to take observations. Therefore, the precise form of the table was difficult to establish and required careful explanation in a footnote. In a case such as this, furthermore, the available climatological data rarely match the engineering requirements; it is a fortunate exception if the nearest mountain-top anemometer station with a long historical record is within 100 km of the proposed telescope site.

5.3. CONTINGENCY TABLES

A *contingency table* displays joint frequencies of occurrence for two or more variables or attributes. It may be *dichotomous,* involving simple yes/no frequencies, e.g., rain or no rain, or it may include class intervals. An example of the latter is given in Table XIII; a hypothetical sample of 297 days has been divided into three subsets of evapotranspiration

TABLE XIII

Hypothetical Contingency Table of Daily Evapotranspiration versus
Daily Solar Radiation, Each Subdivided into Three Subsets

Solar radiation	Daily evapotranspiration			Totals
	Low	Medium	High	
Low	49	30	20	99
Medium	24	44	31	99
High	26	25	48	99
Totals	99	99	99	297

and three subsets of solar radiation ranges. The particular numbers in the table have been chosen to illustrate the fact that biometeorological data rarely yield exact relations; although visual inspection of Table XIII suggests that there is less evapotranspiration on cloudy days, many exceptions occur.*

The impact of a table can sometimes be sharpened by reducing it to a 2-column by 2-row display. Contrasts are further strengthened by using only the upper and lower values of one variable, i.e., by eliminating the central 30–60% of the data, although statistical significance tests cannot then be applied. Tabular and graphical displays suggest but never prove relationships; thus, there is no need to clutter the visual image with too much detail unless a strongly nonlinear trend is to be illustrated. A typical example is given in Table XIV (147). The effect of wind direc-

TABLE XIV

Number of Cases of Very High and Very Low Soiling Index at Ottawa,
Canada in December–January–February, 1956–57
According to Wind Direction[a]

	Wind E, SE, S, SW	Wind W, NW, N, NE
No. of cases of very high soiling index	61	17
No. of cases of very low soiling index	1	51

[a] Munn and Ross (147).

* Alternatively, a table may list average values of a particular variable in each of the various prestratified classes. In this case, the median is often a more appropriate indicator of central tendency than is the mean, if the subsamples are small or are not distributed normally.

tion on air quality in Ottawa is emphasized by considering only small subclasses of extreme values and by separating wind direction into only two groups.

Contingency tables assist in the search for multifactorial relations, particularly when there is a very large sample of data (e.g., hourly observations for a period of at least 15 years) on punched cards or magnetic tape for input to a high-speed computer. Most statistical methods contain assumptions that are rarely satisfied in biometeorology, e.g., the requirements for randomization and linearity. Chisholm and Muller (148) have described a computer data–display system that assumes no physical relations among a number of variables. From a 15-year record of hourly values of visibility, for example, frequency distributions may be prepared for each of a large number of subsets. One such grouping might include only the hours from 3 to 6 AM when the wind direction was from the northeast quadrant, wind speed was 2–4 m/sec, skies were clear, air temperature was between 15 and 20°C, and relative humidity was in the 80–90% range. Provided that there are 100 cases or so in the subset, the table gives a useful visual picture of the combined influence of a number of factors on visibility. This leads to a better understanding of the interactions among the variables.

The computer program designed by Chisholm and Muller is flexible. Class ranges can be changed readily if an initial output shows that the observations are clustered in one subcell. In addition, the variables need not be concurrent; in order to introduce a persistence factor, a contingency table can relate visibility at time $t = 0$ with weather conditions at times $t = -1, -2, -3, \ldots,$ hr.

5.4. GRAPHICAL DISPLAYS

Recent developments in electronic and photographic techniques have created new opportunities for visual synthesis of data. Many of the large computers are equipped with output units which either produce isopleths or display several intensities of shading to indicate gradients on a chart.

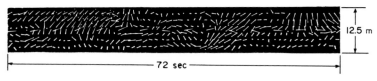

12.5 m

72 sec

FIG. 34. Oscillogram showing wind fluctuation vectors along a cross-wind line, 1600 GMT, May 15, 1963; wind speed 4.5 m/sec. (150).

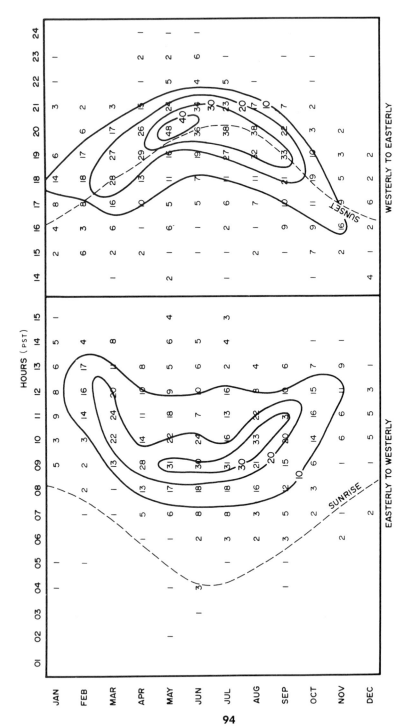

Fig. 35. Frequency of occurrence of diurnal wind shift by hour and month, Vancouver, British Columbia, August 1961–July 1967 (151).

94

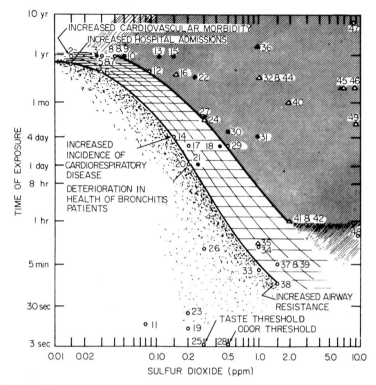

FIG. 36. Effects of sulfur oxides on health; the numbers refer to specific incidents, e.g., the London, England smog of 1952 is represented by point 32 (152); (○) morbidity in man; (●) mortality in man; (△) morbidity in animals; (▲) mortality in animals. The hatched area indicates the range of concentrations and exposure times in which deaths have been reported in excess of normal expectation; the crosshatched area indicates the range of concentrations and exposure times in which significant health effects have been reported; the stippled area indicates the ranges of concentrations and exposure times in which health effects are suspected.

In field investigations too, there are new aids, some of which have been described by Bellamy (149). Jones (150) has given a particular example, namely, the use of a cathode-ray tube to obtain a flow visualization from 6 anemometers all at a height of 2 m over open countryside and aligned at right angles to the wind. One of his photographs (Fig. 34) illustrates the method and indicates the coherency of the flow on that occasion.

Nomograms are graphical representations of the solutions of equations in three or more variables, and they are prepared as time savers. Therefore, there is no great need for simplicity and clarity; the nomogram has been constructed for repetitive use, and the user is expected to acquire a skill, as he would in the case of a slide rule. In most other cases, how-

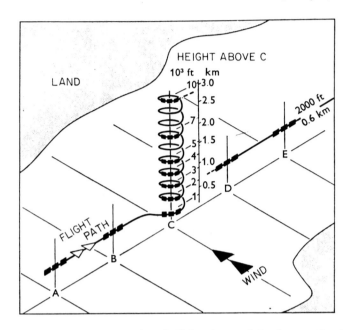

FIG. 37. Diagram illustrating aircraft flight plan used for intercepting immigrant spores over the English Channel; each black rectangle indicates that an air sample was taken at that location (153).

ever, insufficient attention is paid to the question of presenting data in the most meaningful graphical form. As an example, the impact of a contingency table may be sharpened by the construction of isopleths, as illustrated in Fig. 35 (151), a table prepared by Emslie of the frequency of occurrence of diurnal wind shifts at Vancouver, B. C.

With increasing numbers of variables, more and more ingenuity is required. Figure 36 illustrates what can be achieved (152); the results of many investigations concerning the effects of SO_2 on health are compressed into a single diagram. The figure is a vivid summary of the state of knowledge at the time of publication, 1967. Other examples are given in Fig. 10 (Chapter 2) and in Fig. 37, which summarizes the relevant details along the flight path of an aircraft which was sampling for spores over the English Channel (153) (see Section 10.3).

The simultaneous behavior of four or more biometeorological variables cannot be shown in a single diagram. A series of figures must be constructed, one for each season, each time of day, or each wind-speed class, for example. The investigator then has many options in the selection of the particular variable to be kept constant in an individual graph. These options should be carefully assessed.

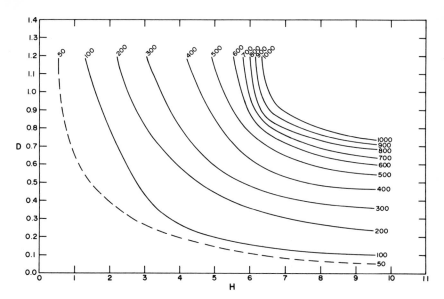

Fig. 38. Isopleths of mass of a hollow cylinder, based on values of height H and density D, drawn from a table of random numbers.

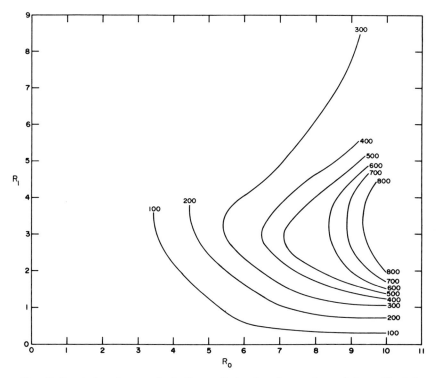

Fig. 39. Isopleths of mass of a hollow cylinder, based on values of the radii of the outer and inner cylinders R_0, R_1, drawn from a table of random numbers.

97

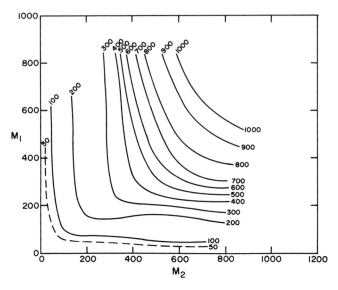

Fig. 40. Isopleths of mass of a hollow cylinder. The abscissa is the best estimate of mass obtained from Fig. 39 while the ordinate is the best estimate of mass obtained from Fig. 38.

A method of multivariate graphical analysis, described, for example, by Panofsky and Brier (45, pp. 180–183), has the merit of not assuming linear relations. The first step is to number the observations in some arbitrary way. For 35 sets of values of 5 variables x_1, x_2, x_3, x_4, and y, say, two initial graphs (Figs. 38 and 39) are drawn with coordinate axes x_1, x_2 and x_3, x_4. On each, the values of y and identification numbers (1–35) are plotted, and smoothed isopleths of y are drawn. Next, a third graph (Fig. 40) is prepared in which the coordinate axes are the smoothed values of y obtained from the first two graphs, and on which a new set of isopleths is drawn. This then gives the best estimate of y for any combination of values of x_1, x_2, x_3, x_4. Finally, the predictions are tested with independent data set aside at the outset for this purpose.

The method is illustrated in Figs. 38–40. The values of x_1–x_4 are taken from sets of random numbers, and are identified as the density and relevant dimensions of a hollow cylinder (154):

$x_1 = H$ = height,

$x_2 = D$ = density,

$x_3 = R_O$ = radius of outer cylinder,

$x_4 = R_I$ = radius of inner cylinder.

Then the predictand y is designated as the mass M, which for comparison can be calculated from the exact relation

$$M = \pi \, DH(R_O{}^2 - R_I{}^2)$$

The predictions were tested with 20 additional sets of random numbers, yielding a correlation coefficient of 0.84 between true values and graphical estimates of M. The largest individual error occurred in a set having R_O almost equal to R_I (6.00 and 5.98), a condition that did not exist in any of the original 35 sets.

The graphical method is sometimes useful when curvilinear relations are suspected or when a computer is not available.

5.5. CARTOGRAPHY AND SPATIAL AVERAGING

Cartography is deceptively simple in theory but infinitely difficult in practice, requiring hundreds of individual judgements by the analyst. There are two main reasons for error:

(a) The network may not be representative of the region, particularly in mountainous country and at island stations.

(b) The data may not be homogeneous, i.e., the observations may not be drawn from a single population. Reasons for this include sensor incompatibility (peak gusts depend on the response characteristics of the anemometer, for example), the presence of nonrandom errors, and the inclusion of observations from different periods of time.

The purpose of a map must be clearly stated, particularly if the cartographer merely wishes to illustrate broad principles such as latitudinal and coastal effects. There is always a risk that the consumer will interpret a map too literally, using it for interpolation between isopleths and for extrapolation into the future or to adjacent areas. An engineer requires a design wind speed or temperature for the lifetime of a structure not yet built, whereas the cartographer is synthesizing historical data. In urban areas where the atlas is most likely to be misinterpreted, large-scale insets should be included so that all the available point observations can be given, with the period of record at each station clearly indicated. Figure 22 is an example of such an inset.

Sometimes the data are too sparse or the terrain too rugged for the analyst to make a judgment on the position of isopleths. The available observations should then be plotted directly on the map, and different colors used to represent high, medium, and low values of the variables. This method has been used in some recent climatic maps of Africa (155).

In a discussion of climatic maps for mountainous regions, Steinhauser (156) has included the following suggestions:

(a) If the isolines are to be used for interpolation, the scale must be sufficiently large that orographic effects can be delineated. Some climatic atlases show isolines crossing hills and valleys in impossible ways. Monthly mean temperatures, for example, are different on north than on south slopes, and there is a discontinuity at the snow line.

(b) An attempt is sometimes made to normalize isotherms for elevation, by using either the adiabatic lapse rate or the actual lapse rate obtained from radiosondes. However, the new values may be misleading, particularly during winter in a closed valley subject to temperature inversions. Furthermore, real isotherms are likely to be more useful for most applications.

(c) For many climatic elements including temperature, an alternative to (b) is to derive a curve of the average height dependence of the element for the entire region; a map is then drawn of the local deviation from the mean for the same elevation. This is a useful method if the network of stations is homogeneous; frequently, however, the height-dependence curve is biased.

(d) Wind roses should be plotted for each station, with separate maps for day and for night because of the alternating upslope and downslope flows.

(e) Insight is required when drawing climatic maps, and the work cannot be done by computers in most cases. In particular, the analyst must be prepared to provide a physical explanation for twists and turns he has drawn in the isopleths. Reproducibility of these features in different seasons or decades lends confidence to their reality.

Kerr *et al.* (157) have described a mesoscale sampling network in Wisconsin for solar radiation. They present spatial comparisons in terms of the percentage of local clear-day global radiation. This normalization decreases the effect of systematic calibration errors and removes the influence of latitude and clear-sky variations of atmospheric aerosols and precipitable water vapor.

Useful advice on climatological mapping is contained in the World Meteorological Organization technical regulations (158). Every effort should be made not to mislead the user. For example, a map of North America showing the frequency of tornadoes is readily misunderstood; if a line labeled 5 tornadoes per 100 km² per 30 years passes through a city, how is it to be interpreted?

Data prestratification is desirable in mapping as it is in many other applications. Figures 41 and 42 (159) show isopleths of total suspended particulates (solid lines) and of iron (dashed lines) in Windsor, Ontario for groups of days when the wind was NNW–NNE (Fig. 41) and when the wind was SSW–SW (Fig. 42); 24-hr periods that included wind shifts have been excluded. The patterns are markedly different in the two subsets of days and for the two types of pollution; a major source of iron is revealed on the far left side of Fig. 42. This kind of insight is not possible from a consideration of monthly or seasonal mean isopleths.

Glasspoole (160) has discussed the estimation of areal total precipitation in a watershed for a single season or year. The first step is to prepare a map of precipitation isopleths (*isohyets*) based on all the data obtained over past decades in and near the area, including stations no longer in operation. Glasspoole notes that when averages are taken over many years, the isohyetal gradients are controlled by orography; thus, reasonable interpolations between observing points can be made from a knowledge of the land features, although it is important to note that an increase in precipitation with increasing height may be reversed near summits. The next step is to compute the ratio (for each station) of the observed short-term rainfall to the long-term climatological value. A spatial average of these ratios is then estimated graphically, from which the watershed total precipitation can be obtained. This is called "the cartographical method" and is based on empirical evidence that the percentage deficiency (or surplus) in any one year is relatively constant throughout a watershed although the annual precipitation varies widely from year to year. In arid zones, the percentages become meaningless and WMO (158) recommends that ratios be replaced by differences. The question of the adjustment of long-term incomplete records to a standard base period is considered in Chapter 7.

The estimation of areal averages of a variable such as precipitation is often done with *Thiessen polygons* (161). Straight lines joining adjacent observing points are drawn on a map; the right bisectors are then constructed, their adjacent intersections forming the vertices of a polygon around each station. Finally, the area of each polygon is estimated with a planimeter and used as a weighting factor to obtain the spatial average of the variable. If the observing points are equidistant, of course, no weighting is required.

In a study of the ecological range of cork oak, in the Mediterranean region, Haggett (162) preferred not to use the Thiessen polygon method to infer spatial distributions of tree populations from observations at fixed points. This is because there were sharp discontinuities rather than modest gradients in the numbers of cork oak. Similarly, the polygon method is inappropriate for the estimation of rainfall in mountain regions.

FIG. 41. Isopleths (μg/m³) of dust loading (solid lines) and of iron (dashed lines) averaged over 11 days when winds were NNW–NNE (Station 3 omitted because of 3 missing days) (159).

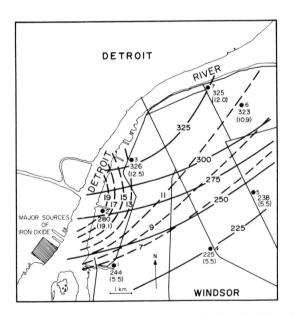

FIG. 42. Isopleths (μg/m³) of dust loading (solid lines) and of iron (dashed lines) averaged over 11 days when winds were SSW–SW (159).

5.6. VECTOR DIAGRAMS

The World Meteorological Organization (158) has recommended standard methods of plotting wind roses. These are illustrated in Fig. 43a for wind directions and Fig. 43b for a combination of speed and direction classes. Other representations are not encouraged.

The direction of the *prevailing wind* is that direction that occurs most often. This quantity is sometimes included in scientific reports but is of little biometeorological value, mainly because biological stresses and the wind both have diurnal cycles. Almost always, but particularly in valleys and along slopes and coastlines, wind patterns are climatologically different at night from those in the daytime. Thus, both a prevailing wind and a monthly or annual wind rose obliterate most of the useful information.* In engineering applications too, conventional wind roses are quite misleading, in the design of structures to withstand gales or driving rain, for example. Conditional frequencies are more meaningful, as Böer (163) and Reidat (164) have emphasized.

Wind roses are often biased. In the first place, the relatively high starting speed of standard anemometers results in an excessive number of indicated calms. This deficiency has been recognized, by Holzworth (165) and Truppi (166) for example, and is important in some air pollution applications. There is no reliable method of removing the bias, although the calms may be distributed according to the frequency of direction in the range just above the anemometer starting speed, as a first approximation. In any event, an analysis of the frequency or duration of calms must never be attempted.

Second, the direction frequencies may be biased. This occurs when winds reported in tens of degrees are converted to 16 compass points. The cardinal directions, N, E, S, and W, are equivalent to 30° each, while the other 12 directions include only 20° each. The 3:2 inequality can be removed arithmetically.

Third, the data may contain observer bias. There is often an apparent greater frequency of winds from the eight primary directions (N, NE, E, SE, S, SW, W, and NW) than of winds from the other eight points. One proposal for removing this effect is as follows (167):

(a) The 16-point scale is transformed to an 8-point wind rose by apportioning, for example, NNE cases to N and NE on the basis of the comparative frequencies of N and NE directions.

---

* National Meteorological Services should be encouraged to publish separate wind frequencies for day and for night.

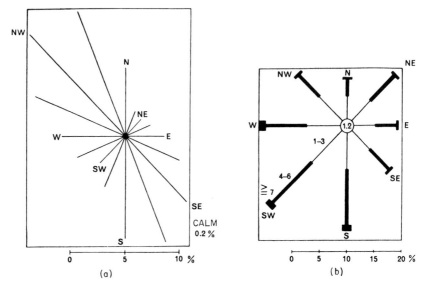

FIG. 43. Recommended ways of plotting wind roses (158): (a) frequency of wind directions for the year, Hurbanova, Czechoslovakia (1946–1955); (b) frequency of wind directions for groups of wind speed (1–3, 4–6, and ≥ 7, Beaufort) for one year, Torslanda, Sweden (1937–1946).

(b) Of the adjusted frequencies of N winds, half are assumed to belong to the central 22.5° of the 45° sector. The other half are then shared between NNW and NNE but weighted according to the comparative frequencies of NW, N, and NE directions.

Because observer bias does not always have the symmetry assumed above, another approach has been suggested by Wallington (168), based on the elliptical frequency distribution. The method is too lengthy to be described here, but one comparison is shown in Table XV for Finke, Australia; the number of cases of NW winds, for example, has been reduced from 147 to 92, with a compensating increase in WNW and NNW winds. Although these corrections seem reasonable, the validity of the method (or any other) cannot be demonstrated rigorously.

Court (169) has suggested that even if wind roses are unbiased, they are nevertheless slightly misleading. In most vector diagrams, the radii are proportional to the frequencies; thus, for example, each line on a 16-point wind rose actually represents a 22½° sector. The alternative is to preserve equal areas rather than equal lengths, a representation that can be achieved easily by making the radii proportional to the square roots of the frequencies. If a central circle of radius c is required for

TABLE XV

WIND DISTRIBUTION FOR FINKE, AUSTRALIA FOR OBSERVATIONS AT 0900 AND
1500 LMT, JUNE–AUGUST, 1957–1963, EXCLUDING CALMS[a]

Directions	Recorded number of observations	Corrected number of observations
NNE	56	61
NE	113	66
ENE	39	64
E	87	71
ESE	36	95
SE	202	119
SSE	67	121
S	170	142
SSW	38	85
SW	81	42
WSW	27	42
W	62	52
WNW	20	67
NW	147	92
NNW	23	65
N	64	52

[a] Wallington (168).

plotting the percentage of calms, the length of each radius as measured from the center is made proportional to $(c^2 + f^2)^{1/2}$, where f is the frequency. Court's suggestion has not been generally adopted. Another representation, the *stereogram*, has been discussed by Brooks and Carruthers (146, p. 168). After distributing the winds into a contingency table of 8–16 directions and 5–10 classes of speeds, one subcell (say, NW, 20–25 km/hr) is taken as reference. The data are then converted to frequency per unit area (frequency density) and isopleths are drawn. These lines may be compared with the Gaussian circular distribution, for example, which consists of concentric circles centered at the apex of the vector mean wind.

Not only winds but sometimes other variables are represented on vector diagrams. The annual cycle of temperature can be portrayed on a 12-point compass (one point for each month) and the diurnal cycle on 24 points (one point for each hour).

Another type of vector diagram, the *phytograph*, is used in ecology. Four selected indices of species performance within a stand are normalized to relative values; these indices usually include relative density, relative basal area, relative cover, and one of relative biomass, productivity, or volume. The four values are then drawn on a four-point

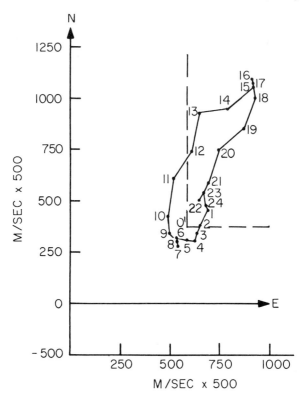

Fɪɢ. 44. The 24-hr average sea-breeze hodograph for Halifax, Nova Scotia in July. The numbers 1 to 24 within the diagram are the hours on a 24-hr clock ᴀsᴛ. The inner coordinate axes (dashed lines) are centered on an origin that is the estimated position of the vertex of the vector mean monthly wind resulting from macroscale influences alone (171).

compass, and the ends of the radii are joined. Although the enclosed quadrilateral is supposed to represent an integrated mass effect of four variables in some way, the term "mass effect" is not defined. McCormick and Harcombe (170) have surveyed the literature and have suggested independently of Court (169) that the four radii be made proportional to the square roots of the frequencies.

A *hodograph* is the locus of one end of a variable vector as the other end remains fixed; the diagram is used to depict the diurnal cycle of surface wind, or the variations of wind with height. An example is given in Fig. 44, which shows the 24-hr sea-breeze hodograph for Halifax, Nova Scotia in July (171).

Wind variations in both time and three-dimensional space are difficult to portray. Compromises are awkward and require the cross referencing of a number of figures by the user. The vector wind may be separated into its orthogonal north–south and east–west components; alternatively, arrows can be used, with barbs or numbers indicating speeds. The principal axis is frequently oriented along a valley or a coastline for mesoscale analysis.

The variation of an element such as air quality or evapotranspiration rate is often dependent on wind, and there is need for a generalized biometeorological wind rose. Radial distance may be used to indicate, for example, the average concentration of SO_2 for a given wind direction. Alternatively, the vector may represent the percentage of time that SO_2 concentrations exceed some value. In the latter case, the percentage may be low simply because the wind rarely blows from that direction; thus a preferred indicator is the conditional frequency, i.e., the percentage of the time that concentrations are high, given that the wind is in the specified direction. For example, easterly winds may be infrequent but when they do blow, air quality is poor.

When the behavior of a biometeorological variable is to be depicted in relation to both wind speed and direction, the representation can be of the form illustrated in Fig. 45. The diagram shows isopleths of SO_2 concentrations for eight wind directions and a number of wind-speed classes. The data are for a particular sampling station in Ottawa, and the effect of individual point sources is evident. This kind of analysis requires prestratification according to the time of day and season. Because winds are usually less at night than during the day, a 24-hr wind rose is biased, and night-time observations are clustered near the origin. The bias is even more serious at locations subject to land/sea or upslope/downslope diurnal wind reversals.

A contingency wind rose has been discussed by Crutcher (172). Instead of plotting mean values of a variable on a diagram such as Fig. 45, the number of occurrences of high SO_2 values, of precipitation, or of fog is plotted for each subset; isopleths are then constructed.

For climatological comparison of winds at two locations, Untersteiner (173) has developed the graphical method illustrated in Fig. 46. The reference frame is the wind at one station (A). Values of the direction and speed at another station (B) are plotted for given simultaneous values of A, usually averaged over each of a number of class intervals. Lines of equal speed (*isotachs*) and of equal direction (*isogons*) are then drawn. In Fig. 46, for example, when the wind at Station A is west 20 km/hr, the most likely wind at Station B is westsouthwest 25 km/hr.

Alternatively, tabular displays are used. Cehak (174) suggests a matrix of the form illustrated in Table XVI, from which four parameters may be derived:

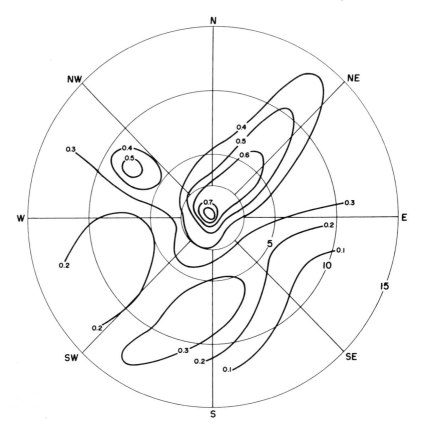

Fig. 45. Pollution wind rose for a sampling station in Ottawa, Canada, based on daytime hourly observations (0900–1700 EST) during the months November–April and the years 1961–1965. The central point in the diagram represents calm winds while the concentric circles denote speeds of 2.5, 5, 10, and 15 m/sec. The superimposed isopleths represent mean concentrations of SO_2 in parts per ten million. Winds were obtained from an anemometer at a height of 60 m in the city.

(a) S: The percentage of winds simultaneously blowing from the same direction at the two stations (obtained by summing the diagonal frequencies).

(b) SD: The sum of the elements in the two boxes just above and in the two boxes just below the main diagonal; this is a measure of directional differences within $\pm 90°$.

(c) D: The sum of the elements in the two boxes just above minus the sum of the elements in the two boxes just below the main diagonal; this is a measure of the turning of the wind.

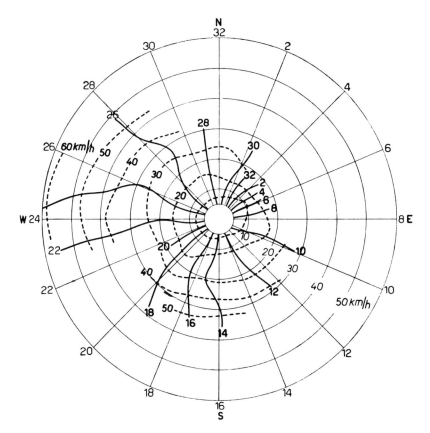

FIG. 46. Isotachs of wind speed (dashed lines) and isogons of wind direction (solid lines) at Station B associated with given wind speed and direction at Station A (reference frame) (173).

(d) *R:* The residual sum of all elements more than two boxes away from the main diagonal.

Cehak suggests that these four parameters be used in cartography. Particularly for mesometeorological networks of stations, the construction of isopleths of S, SD, D, and R is recommended in preference to a series of wind roses.

Finally, Fig. 47 (175, p. 54) is an example of a simple but effective way of depicting pollution wind roses (or of other biometeorological variables whose values are influenced by wind direction). The diagram suggests but, of course, does not prove that a major source of SO₂ is situated between the sampling stations.

TABLE XVI

CONTINGENCY TABLE OF CONCURRENT WIND DIRECTIONS AT STATIONS A AND B, FOR WIND SPEEDS $\geq$10 KM/HR AT STATION A; THE NUMBER OF OBSERVATIONS WAS 3907[a,b,c]

B \ A	NNE-NE	ENE-E	ESE-SE	SSE-S	SSW-SW	WSW-W	WNW-NW	NNW-N	Sum
NNE-NE	8	0	0	1	1	0	0	2	12
ENE-E	0	3	1	3	0	0	0	1	8
ESE-SE	0	3	63	19	0	1	0	0	86
SSE-S	0	0	17	81	0	0	0	0	98
SSW-SW	0	0	0	3	2	5	0	2	10
WSW-W	3	0	2	4	2	341	76	37	430
WNW-NW	0	2	0	1	0	62	176	51	278
NNW-N	4	0	1	0	0	8	14		78
Sum	15	8	84	112	5	417	266	93	1000

[a] Cehak (174).
[b] S = 725, SD = 263, D = 31, R = 12 (see text for definition).
[c] The table has been normalized to 1000 observations.

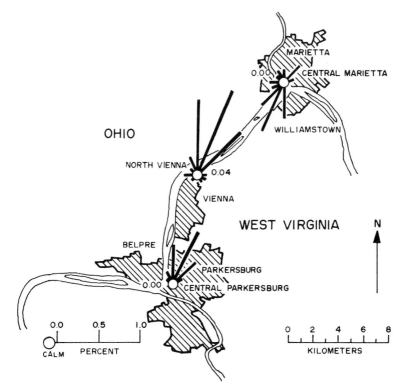

FIG. 47. SO₂ pollution roses for concentrations >0.10 ppm, October 1965 through February 1966 in the Parkersburg, West Virginia–Marietta, Ohio area (175).

5.7. STREAMLINES AND TRAJECTORIES

At any instant in time, a two-dimensional wind flow can be specified by isotachs and by *streamlines* (a streamline is a line whose tangent at any point is parallel to the instantaneous velocity of the fluid at that point). Streamline analysis is an important tool in synoptic forecasting in the tropics; the method is of value also for delineating mesoscale flows.

The procedure is as follows (176, p. 23). The analyst first constructs isogons (lines of equal wind direction) at intervals of 30° (or 10° in regions of small gradients); he then draws short straight-line segments, aligned in the direction of the wind, at a large number of points equally spaced along each isogon; finally, he connects the line segments by tangent curves, the streamlines. An example of an isotach-streamline analysis is shown in Fig. 48, a diagram prepared by Schultz and Fitz-

FIG. 48. Streamlines (dark lines) and isotachs (light lines) delineating the meso-scale wind flow in the Sacramento Valley of California on a July day between 8 and 9 AM (177).

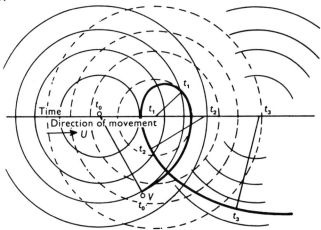

FIG. 49. Cycloidal trajectory (heavy solid line) of a fluid element, in response to the eastward motion of a center of low pressure (178).

water (177); the mesoscale flow in the Sacramento Valley of California is revealed in some detail.

The *trajectory* of a fluid element is the curve described by successive positions of the element. This curve is often of more importance than the streamlines, particularly in the transport of a radioactive cloud and in the spread of pollen and spores. The trajectories and streamlines coincide only in steady-state conditions (no accelerations). With a moving depression or anticyclone, the trajectories of individual elements may not be at all obvious from inspection of a series of synoptic weather maps.

Trajectories are constructed from consecutive synoptic charts for times t_1 and t_2 (176, p. 27), as in the following example. Suppose that a small cloud of pollen is at position P_1 at time t_1. The vector wind at P_1 is used to obtain a simple linear extrapolation of the movement of the cloud to position P_2 at time t_2. The new position P_2 is transposed to the next synoptic map, and the vector wind at that point (at time t_2) is used to extrapolate the motion backward in time to t_1. Under steady-state conditions, the resulting location would be P_1; in most cases, however, this is not so, and an average of the two displacement estimates is taken. Finally, the procedure is repeated for subsequent times t_3, t_4 . . . .

Hirst and Hurst (178) have given a number of examples of trajectory analysis; for the particular case of an eastward-moving depression illustrated in Fig. 49, fluid elements travel along cycloidal loops. The finite-difference process and uncertainties in the wind estimates naturally cause some errors in the trajectories. Sometimes also, the initial height of the tracer cloud may not be known while subsequent heights may be very uncertain because of organized vertical motions in the atmosphere: in such cases, the predicted trajectories may be seriously in error after only a few hours. An estimate of the cumulative effect of all these uncertainties has not yet been made but studies are in progress at the University of Stockholm (H. Rodhe, personal communication). Assuming horizontal flow, Durst and Davis (179) have suggested a climatological average error of 13 km/hr (7 kt) in the wind estimates obtained from upper-air synoptic maps.

6/STATISTICAL METHODS I

6.1. USES OF STATISTICAL METHODS

In this chapter and the following one on time series, familiarity with elementary probability and statistical theory is assumed, there being many textbooks on the subject. The texts by Brooks and Carruthers (146) and by Panofsky and Brier (45) are recommended because of their meteorological emphasis. To begin, however, the strengths and weaknesses of the classical statistical methods should be mentioned.

First, statistical parameters are of value in *summarizing* the properties of a finite population. Instead of listing all the observations, the investigator may give an average, one or more indicators of variability, and possibly a frequency distribution or the parameters of a mathematical model of that distribution. An example of a finite population is all hourly temperatures observed in July 1969 at a single site.

Second, statistical methods are of value in deciding whether a subset can reasonably be assumed to belong to a larger population, in deciding whether two subsets have been drawn from the same population, and in inferring the properties of a population from those of one or more subsets.

Third, statistical methods are useful in determining whether or not there is an association among two or more variables. No inferences can be made, of course, about cause and effect. For example, photosynthesis and transpiration rates are correlated because each is correlated with other environmental factors such as solar radiation flux, temperature, and soil nitrogen deficiency. Many interpretive difficulties of this type occur in biometeorology.

The classical methods require *unbiased* samples, i.e., the members of the subset must be obtained in a completely random manner. The July

114

1969 temperatures do not meet this criterion; each hourly value has a "memory" and is correlated with the temperatures of the previous few hours. In fact, there are few geophysical data that do not display time dependence among the members of any subset.

Statistical methods are invaluable when unbiased observations are available. Analysis of variance, Latin squares, chi square, t, and F tests assist in deciding whether to accept, in the presence of experimental noise, the reality of similarities or differences and of associations. In some cases too, the number of observations required to obtain a significant result may be determined. Statistical methods are particularly appropriate in replication experiments. An air pollution application has been described by Sanderson *et al.* (180), a comparison of the efficiency of five types of dustfall collectors, using a replicated Latin square arrangement on the roof of a building.

6.2. STATISTICAL NORMALIZATION

Normalization is a recurring theme in this book, and a few comments are not out of place here. Normalization in the statistical sense is the replacement of a variable X by a transformed value z:

$$(6.1) \qquad z = (X - \bar{X})/\sigma$$

where $\bar{X}$ and σ are the mean and standard deviation, respectively, of the population of X. The quantity z is said to be located at $\bar{X}$ and scaled in σ; z is dimensionless, whereas X may be a length, temperature, mass, etc.

When two variables X and Y are compared, they may both be in normalized form. This procedure is particularly appropriate when X and Y are to be plotted on a single ordinate scale, the objective being to show both as functions of a common third variable Z. Alternatively, Robinson and Bryson (181) suggest that the linear regression $Y = aX + b$ be computed and that the individual departures of Y from the predicted value be examined as functions of Z. For example, isopleths of anomalies may reveal distinctive spatial patterns, inviting subsequent consideration of other factors.

A normalized index frequently encountered in biometeorology is the *coefficient of variation* C_V. For a Gaussian distribution,

$$(6.2) \qquad C_V = 100\sigma/\bar{X}$$

The quantity X must be measured on an absolute scale (146, p. 41), i.e., the index is not relevant when temperature is in degrees Fahrenheit or degrees Celsius. For a log-normal distribution with geometric standard deviation s_g,

(6.3) $C_V = 100 \text{ antilog } s_g - 100$

When values of X are drawn from a network of randomly spaced stations, the expression in Eq. (6.2) or (6.3) is called the *coefficient of geographic variation* C_{GV}. This index has been used by Stalker *et al.* (76), for example, to determine the minimum number of pollution sampling stations required to estimate the true areal mean within specified confidence limits.

(6.4) $n = (C_{GV})^2 t^2 / p^2$

where

    $n =$ the minimum number of stations,
    $t =$ Student's t for specified confidence,
    $p =$ the allowable percentage departure from the true mean.

6.3. EMPIRICAL STATISTICAL METHODS

Given a set of nonrandom data, a mean and standard deviation can, of course, be computed. These parameters summarize two of the properties of the set although they cannot be used to estimate the probability that the set is drawn from some larger population. In many cases, nevertheless, several subsets have means and standard deviations that change little from place to place or year to year. For example, if the average July temperature over the last 50 years at a particular location has ranged from 17 to 25°C, with 90% of the values between 19 and 23°C, a reasonable prediction is that next year's value will be in the range 19–23°C, with an empirical probability of about 90%.

Correlation coefficients can be computed for nonrandom observations of two variables but again, significance tests cannot be used. Panofsky (182) has considered the correlation that might exist between hourly values of pressure and temperature in New York City in a particular March. If anticyclones from the Canadian arctic dominated the weather, a correlation coefficient of -0.5 might be obtained; with the unbiased-sample assumption (not justified, as Panofsky emphasizes), the result would be significant at the 1% level. In another year, on the other hand, March might be a period of warm weather associated with subtropical anticyclones, yielding a correlation coefficient of $+0.5$. Each of the correlation coefficients provides information about the meteorological processes within the particular time period but is of no value as a general predictor of March temperatures from hourly values of pressure.

Brooks and Carruthers (146, p. 44) have illustrated the influence of nonrandomness by considering a time series (containing no trend) in which there is a correlation coefficient r between successive observations. For $r = 0$, as is well known,

(6.5)
$$\sigma' = \sigma/\sqrt{2}$$

where

$\sigma' =$ the standard deviation of the distribution of mean values of a number of samples, each containing 2 observations,

$\sigma =$ the standard deviation of the parent population.

In the presence of a correlation r, however, the standard deviations of means of successive pairs of observations is not $\sigma/\sqrt{2}$ but $\sigma[(1 + r)/2]^{1/2}$.

In another special case (146, p. 227), if X and Y have equal variances but are uncorrelated, it can be shown that the correlation between $(X + Y)$ and X is 0.7. Similarly, if X, Y, Z are uncorrelated but have the same coefficient of variation, the correlation between X/Y and Z/Y is about 0.5. Many subtle variations of these rather trivial examples exist in biometeorology.

Frequency distributions of nonrandom samples are of interest, particularly in the estimation of empirical probabilities that certain values will or will not be exceeded. Applications include studies of air quality criteria, wind loadings on buildings, and lake levels. A Gaussian distribution is to be expected when there are many causal factors, each having a small independent effect on the variable. When the influence of the factors is multiplicative, on the other hand, Yevdjevich (183) suggests that the data are distributed log normally. (See Section 7.8 for a further discussion of this point.) In any event, this latter distribution is encountered frequently in biometeorology. Its parameters are the geometric mean or median and the geometric standard deviation. Schubert *et al.* (184) find that the strontium-90 concentrations in human beings can be fitted a log-normal curve. Although the median value varies in different areas of the world, within particular age groups, and for different time periods, the geometric standard deviation seems to change little; thus, about the same proportion of a particular population contains concentrations in excess of a given multiple of the median.

Zimmer and Larsen (185) have noted that concentrations of air pollution at urban sampling stations are distributed log normally within the cumulative frequency ranges from 5 to 95%. They therefore recommend that air quality criteria be established on the basis of geometric means (or medians) and geometric standard deviations. A typical distribution is shown in Fig. 50.

A rather different proposal has been made by Stratmann and Rosin (186). They suggest that because pollution distributions are skewed

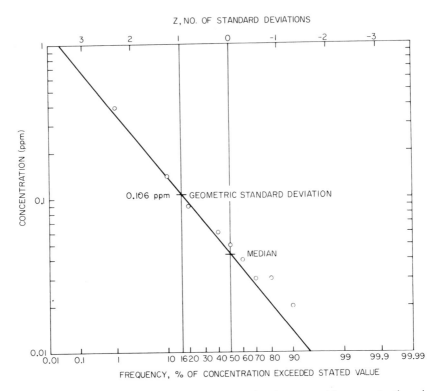

FIG. 50. Frequency distribution of mean hourly nitrogen oxide concentrations in Washington, D. C. from December 1, 1961 to December 1, 1964 (185). $s_g = 16$ percentile/50 percentile $= 0.106$ ppm/0.043 ppm $= 2.46$; $m_g = 0.043$ ppm.

rather than Gaussian, the arithmetic standard deviation s should be replaced by an empirical one s_0:

$$(6.6) \qquad s_0 = \{[2\Sigma(c - \bar{c})^2]/(2n - 1)\}^{1/2}, \qquad c > \bar{c}$$

where n is the number of values of c greater than the mean $\bar{c}$. For 20 SO_2 sampling stations in Duisburg, Germany, the ratio s_0/s ranged from 1.27 to 1.59 with an average value of 1.44 ($s_0/s = 1$ in a Gaussian distribution). The quantity s_0 is recommended by Stratmann and Rosin as an index of air quality.

Juda (187) discusses the two approaches described above. He, however, concludes that the best indicator of air quality is the percentage of time that designated concentration levels are exceeded. The diversity of the various proposals suggests that no empirical representation is completely satisfactory.

Lundqvist (188) considers the response to the perception of odors by an individual over successive trials or by a group of subjects on one occasion. He suggests fitting the cumulative frequencies to an analytical form:

(6.7) $$R_c = \tfrac{1}{2}\tanh(c \cdot a) + 50$$

where

R_c = the percentage response to a given concentration c,

a = a parameter whose value is constant in any one individual or in a group of people but which changes with sex, age, health, etc.

The number 50 is included to make the equation symmetric about the midpoint of the cumulative frequency distribution.

6.4. MULTIVARIATE ANALYSIS

From a table of random numbers, Wallis (154) generated a set of 75 hollow cylinders, each having a given height H, outside radius R_O, inside radius R_I $(R_O > R_I)$, and density D. The values of the derived quantities listed in Table XVII were then calculated.

TABLE XVII

VARIABLES AND THEIR SYMBOLS FOR THE HOLLOW-CYLINDER PROBLEM[a]

Variable	Symbol	Function
—	K	Constant (π)
1	H	Height
2	HH	(Height)^2
3	$2KR_OH$	Outside curved surface
4	$2KR_IH$	Inside curved surface
5	D	Density
6	DD	(Density)^2
7	DG_O	Density times diagonal of outside cylinder
8	DG_I	Density times diagonal of inside cylinder
9	R_O	Radius of outside cylinder
10	KR_OR_O	End area of outside cylinder
11	G_O	Diagonal of outside cylinder
12	R_I	Radius of inside cylinder
13	KR_IR_I	End area of inside cylinder
14	G_I	Diagonal of inside cylinder
15	M	Mass

[a] Wallis (154).

After setting aside the data from 15 cylinders for subsequent independent teting, the best estimate of the mass M was obtained by multivariate analysis methods using the 14 variables listed in Table XVII as predictors. Linear models were tested of the form

$$(6.8) \qquad y = a_0 + a_1x_1 + a_2x_2 + \cdots + a_{14}x_{14}$$

Exact relations cannot be expected because of the nonlinear form of the correct solution:

$$(6.9) \qquad M = \pi DH(R_O^2 - R_I^2)$$

Furthermore, the predictors are intercorrelated, i.e., too many variables have been used to describe the system. This is a situation encountered frequently in biometeorology, but the effect of overspecification is then difficult to isolate because of the added complications of experimental error and bias. Wallis' methodology has widespread applicability.*

The resulting best-fit solution for Eq. (6.8) is given in Eq. (6.10), along with values of Student's t for each term; a significant correlation exists when t is greater than 2.0.

$$
\begin{aligned}
(6.10) \quad M = &+956.0 + 89.7(H) + 7.82(HH) + 4.55(2KR_OH) \\
& \quad\ t = 1.0 \qquad t = 1.7 \qquad\ t = 2.8 \\
& - 7.22(2KR_IH) - 1550.(D) + 589.0(DD) \\
& \quad t = 5.9 \qquad\quad t = 2.8 \qquad t = 2.1 \\
& + 90.3(DG_O) + 3.30(DG_I) - 334.0(R_O) \\
& \quad t = 3.4 \qquad\qquad t = 0.2 \qquad\qquad t = -1.2 \\
& + 2.94(KR_OR_O) + 57.6(G_O) + 595.0(R_I) \\
& \quad t = 1.3 \qquad\qquad t = 0.4 \qquad\quad t = 4.8 \\
& + 1.78(KR_IR_I) - 312(G_I) \\
& \quad t = 0.7 \qquad\quad t = -3.7
\end{aligned}
$$

The coefficient of multiple determination is high: 0.92. Nevertheless, the coefficients are difficult to interpret and provide little information about the underlying physical relation. For example, the variables D and D^2 are both significantly correlated with mass but their coefficients have opposite signs (-1550.0 and $+589.0$). In actual fact, positive coefficients would be expected in both cases.

The effect of removing one of the variables from the analysis is illustrated in Eq. (6.11); the term $a_{12}R_I$ has been excluded.

* As a trivial but nevertheless important additional example, computer programs should be tested with dummy data that yield known values.

(6.11) $M = 763.8 + 17.3(H) + 2.96(HH) + 1.64(2KR_0H)$
$\qquad\qquad t = 0.2 \qquad t = 0.5 \qquad t = 0.90$
$\qquad - 1.76(2KR_IH) - 1486.(D) + 277.1(DD)$
$\qquad\quad t = 3.0 \qquad\qquad t = 2.2 \qquad t = 0.9$
$\qquad + 114.2(DG_0) - 7.12(DG_I) + 142.0(R_0)$
$\qquad\quad t = 3.6 \qquad\qquad t = 0.3 \qquad\qquad t = 0.5$
$\qquad + 5.28(KR_0R_0) - 218.1(G_0) - 5.52(KR_IR_I)$
$\qquad\quad t = 1.9 \qquad\qquad t = 1.2 \qquad\qquad t = 2.6$
$\qquad + 77.7(G_I)$
$\qquad\quad t = 2.7$

Comparison of Eq. (6.11) with (6.10) shows that the coefficients are unstable; for example, the coefficient of G_I has changed from -312 to $+77.7$.

Finally, when only the four correct variables are used, the resulting Eq. (6.12) is more meaningful and the coefficients have the correct signs:

(6.12) $M = -783. + 74.6(H) + 633.(D) + 3.18(KR_0R_0)$
$\qquad\qquad t = 7.2 \qquad t = 5.7 \qquad t = 8.1$
$\qquad - 3.38(KR_IR_I)$
$\qquad\quad t = -6.3$

However, the coefficient of multiple determination has diminished from 0.92 to 0.72, and in a real situation furthermore, the investigator might not know which variables to exclude.

A second multivariate method is stepwise multiple regression. Each of a group of variables is examined in turn; a 95% significance level is necessary for a variable to be included in the final regression while a 90% significance level is required for exclusion at each step. Equation (6.13) shows the results for the hollow-cylinder data:

(6.13) $M = -310 + 2.78(2KR_0H) - 2.70(2KR_IH) + 37.9(DG_0)$
$\qquad\qquad t = 12.0 \qquad\qquad t = 9.2 \qquad\qquad t = 7.6$

The coefficient of multiple determination is 0.82 but the relation is again difficult to interpret physically.

Wallis includes comparative results obtained from four other multivariate methods: principal component analysis, varimax rotation of a factor weight matrix, key cluster regression, and object analysis (see Wallis' paper and his references for descriptions of these methods). When the equations were tested with independent data from the other 15 hollow cylinders, the correlations between predicted and observed masses were between 0.68 and 0.77, with the exception of 0.50 for the object-analysis method.

When using multivariate methods in real situations, the predictors are often intercorrelated. Second, the samples may not be random, and significance tests are then meaningless. Third, there are experimental errors in the independent variable as well as in the predictors; a regression analysis assigns all the errors to the dependent ones. Finally, the physical relationships may be strongly nonlinear; for example, suspended particulate pollution tends to be high with light winds (stagnation situations) and with strong winds (reentrainment of dirt from the ground); the strong-wind maximum is absent, however, when the underlying surface is damp or snow covered. For all these reasons then, multivariate analysis should only be used if no other method is available. Wallis' hollow-cylinder experiment is a reminder of the failure of the method to provide any physical understanding of the nature of a relation. Anderson (189) in fact states that if multivariate analyses are not based on accurate and complete data and/or a physical model, "they are bound to be contrived nonsense."

In a number of environmental applications, the predictors may be represented by "power laws" over usefully broad ranges. Nonlinearity may therefore be removed by taking logarithms. One engineering example is related to the development of safety procedures at Cape Kennedy and Vandenberg AFB, California. On exposure to the atmosphere, the propellants of the Titan II missile emit toxic vapors. Nou (190) has developed on multivariate regression for diffusion at these two launch sites based on a power-law assumption.

$$(6.14) \qquad\qquad C/Q = K x^a \bar{u}^b \sigma_A{}^c (\Delta T + 10)^d$$

where

C = the downwind center-line concentration,
Q = source strength,
x = downwind distance,
$\bar{u}$ = the mean wind speed,
σ_A = the standard deviation of the azimuth angle of the wind direction,
ΔT = the vertical temperature difference (negative during lapse and positive during inversion conditions),
10 = a number added to ΔT to avoid the possibility of having to raise a negative value to a power,
K, a, b, c, d = parameters of fit to be determined by regression methods.

Equation (6.14) would be accepted by most diffusion meteorologists as a reasonable empirical model for modest travel distances. The equation may be converted to linear form by taking logarithms:

$$(6.15) \quad \log C/Q = \log K + a \log x + b \log \bar{u} + c \log \sigma_\mathrm{A} + d \log (\Delta T + 10)$$

The coefficients were determined from measurements made during a large number of diffusion trials at the two sites and at O'Neill, Nebraska. Different data prestratifications led to different coefficients but the final equation developed for day-to-day use was

$$(6.16) \qquad C/Q = 0.00211 x^{-1.96} \sigma_\mathrm{A}^{-0.506} (\Delta T + 10)^{4.33}$$

Note that the mean wind speed is absent in Eq. (6.16). The stepwise regression technique indicated that its inclusion did not reduce the variance significantly. Although this is not a physically satisfying result, presumably the influence of the wind is included empirically in the values of σ_A and ΔT.

The reliability of prediction equations is usually expressed by the percentage reduction of error (variance) of predicted values Y' with respect to observed values Y:

$$(6.17) \quad \% \text{ reduction of error} = 100 \left[1 - \left(\frac{\text{RMS error of } Y'}{\text{standard deviation of } Y} \right)^2 \right]$$

This was not a suitable index in the present case because C/Q varied over six orders of magnitude and because power law Eq. (6.14) had been transformed to linear form, Eq. (6.15), and back to power law, Eq. (6.16); Eq. (6.17) proved to be sensitive to a very few large values of Y. A better engineering "yardstick" was therefore chosen, the percentage of cases that fell within a factor of 2, i.e., within 50–200% of the observed values.

Equation (6.16) was developed using only half the data from each of the three sites. When it was tested with the remaining data, 72% of the cases were within a factor of 2 of the observed values, while 97% were within a factor of 4. This example illustrates that multivariate analysis can be useful in engineering studies, viz., in permitting predictions to be made with a specified probability of success, provided that the population is homogeneous.

Instead of multivariate analysis, it is sometimes necessary or desirable to use discriminant analysis. In the latter method, the predictand and some or all of the predictors may assume only a finite number of values. The simplest example is two-group dichotomous alternatives (rain or no rain, hay fever or no hay fever). Additionally, when a variable has a very wide range of values, the presence of an extreme in a small sample may make the coefficients unstable in a standard regression analysis. Then variable may be prestratified into a few classes such as high, medium, and low. The straight line of best fit discriminates among the alternatives. A useful description of the method has been given by Panofsky and Brier (45, p. 118).

6.5. SPATIAL RELATIONS

Given any set of functions $u_1(x)$, $u_2(x)$, . . . , $u_n(x)$ defined and square integrable for $a \leq x \leq b$ such that

$$(6.18) \qquad \int_a^b u_m(x) \cdot u_n(x) \, dx = \begin{cases} 0 & \text{for } m \neq n \\ \alpha_n^2 & \text{for } m = n \end{cases}$$

the set $u_n(x)$ is said to be orthogonal on $[a, b]$; furthermore, if $\alpha_n = 1$, the set is *orthonormal*.

As a simple example, it can be shown that

$$(6.19) \qquad \int_0^{2\pi} \cos m\theta \cos n\theta \, d\theta = \begin{cases} 0 & \text{for } m \neq n \\ \pi & \text{for } m = n \end{cases}$$

The functions $u_i(x)$ can be considered as vectors in n-dimensional space. The integral in Eq. (6.18) is proportional to the covariance of u_m and u_n; by definition then, orthogonal functions are uncorrelated. Note also that α_n^2 is the variance of the function $u_n(x)$ over the interval a to b.

A geophysical field can be represented by a set of orthogonal functions. Because these functions are independent, the contribution of each to the total variance of the field is also independent. As a well-known illustration, harmonic analysis (146, Chapter 18) is used to separate a diurnal cycle (or an annual cycle) into a finite number of harmonics. More generally, Tchebycheff polynomials (45, p. 169) and eigenvectors (as described below) have been employed to dissect space fields. A primary objective is to synthesize large quantities of data into a much smaller number of parameters that still retain all the essential information of the original field. (Because observations from adjacent stations are correlated, for example, the original data field contains redundant information.) A second objective is to provide a physical interpretation; claims are made that the eigenvector patterns can be "explained," but this is a rather controversial proposition. The eigenvector method is also called *factor analysis*.

The eigenvector approach has been used, for example, by Stidd (191) in a study of monthly precipitation patterns from 60 stations in Nevada, by Sellers (192) in an examination of precipitation over the western

United States, and by Grimmer (193) in a study of 30-day surface temperature anomalies at 32 stations in Europe and the eastern North Atlantic, using 80 years of records. These latter data can be represented for a particular calendar month by a matrix consisting of m vectors each containing n elements, where $m = 80$ and $n = 32$. The objective is to find a representation in the following linear form:

$$(6.20) \qquad F_j = \sum_{k=1}^{32} b_{kj} P_k, \qquad j = 1, 2, \ldots, 80$$

where F_j is the vector field in the jth year for all i, P_k is an orthogonal vector set of patterns occurring in linear combination, and b_{kj} are empirical coefficients. Because Tchebycheff polynomials and other ordinary mathematical functions predetermine the form of P_k, there is merit in letting the data themselves provide a basis for the most appropriate representation. This is particularly useful if nothing is known in advance about the existence or nature of constituent patterns, or if the objective is to find patterns such that the coefficients of only a few of them contain most of the information in the original fields. A "most appropriate" set might have the following properties:

(a) Members of the set of functions are in descending order according to the magnitude of the fraction of the variance that they explain in the original data.

(b) The coefficients b_{kj} are themselves orthogonal over the set of cases $j = 1, 2, 3, \ldots, n$.

Property (a) implies that the P_k are more *efficient* than any other set of functions in representing the data. They are more *efficient* in the sense that any subset of the more important P_k represents the variance of the original data better than does a subset of a like number of functions from any other parent set of orthogonal functions.

Property (b) implies that the n values of b_{kj}, $k = 1, 2, \ldots, n$, which characterize each case j, are not correlated, in contrast to the original data.

It can be shown that: (a) these two properties are not independent, so that to require one is to require the other, (b) within a multiplicative constant, the conditions imply a unique set of functions, and (c) such a set of functions can always be found from sets of observed data.

A method of solving Eq. (6.20) has been described by Grimmer (193), Sellers (192), and Stidd (191), among others. As Stidd indicates, a high-speed computer is almost mandatory, although the method was known as early as 1846.

For Grimmer's European temperature data, the most important pattern accounted for 44% of the total variance in January; the first eight patterns cumulatively "explained" 88% of the variance.

A difficulty with eigenvector analysis is that the derived patterns are not observable directly (194), making physical interpretation uncertain. A preferred approach is to apply inductive reasoning. Years in which particular patterns explain a larger than average fraction of the variance are examined separately to see what anomalies, if any, exist in the associated pressure or wind fields.

6.6 EXTREME-VALUE ANALYSIS

Consecutive measurements of an environmental variable are usually correlated. For example, maximum temperatures on successive days show persistence on the average because of the influence of large-scale weather patterns. These time series can be studied (see Chapter 7); alternatively, a series of extremes consisting, say, of the highest daily value in each year may be examined by more conventional methods provided that there is no serial correlation in the set.

Water levels in rivers usually rise to a maximum at about the same time every year. Provided that there is no trend, or abrupt change due to construction of a dam, the annual peak may be a suitable variable for extreme-value analysis. The lowest annual water level in a river, on the other hand, could occur conceivably on December 31 or January 1; the same extreme condition might then be counted twice in successive years. This difficulty is overcome by commencing each 12-month period at some other time of year. In both cases, however, a test for serial correlation should be performed.

Trend must be removed from a time series before beginning an extreme-value analysis. If a temperature-observing site has become increasingly urban during a 50-year period, the chances of extreme cold have lessened, thus biasing predictions from past observations. Methods of trend removal are considered in Chapter 7.

The description of extreme-value theory given by Kendall (195) will be followed. For convenience, the set of extremes is assumed to consist of n annual maxima or minima, $X_1, X_2, X_3, \ldots, X_n$, although any other time interval ensuring independence could be used. In the case of maxima, let X be some chosen value of X_i. Then if p is the probability of occurrence of $X_i \geq X$, and q is the probability of nonoccurrence $(p + q = 1)$, the *return period* T (in units of years) is defined as

(6.21) $$T = 1/p$$

TABLE XVIII

PROBABILITY P_T THAT AN EVENT WITH RETURN PERIOD T WILL OCCUR
WITHIN THE NEXT T YEARS[a]

T:	2	5	10	20	50	100	200	500	1000
P_T:	0.75	0.67	0.65	0.64	0.64	0.63	0.63	0.63	0.63

[a] Kendall (195).

The probability that X will occur for the first time in the rth year is $q^{r-1}p$. Thus, the probability P_r that X will occur at least once in the next r years is the sum of the probabilities of its occurrence in the first, second, . . . , rth years.

(6.22) $$P_r = p + pq + pq^2 + \cdots + pq^{r-1}$$

substituting $p = 1 - q$,

(6.23) $$P_r = 1 - q^r = 1 - (1 - 1/T)^r$$

For the special case $r = T$

(6.24) $$P_T = 1 - (1 - 1/T)^T$$

Equation (6.24) is tabulated in Table XVIII (195). For a dam designed to withstand a flood with a 50-year return period, there is a 64% chance that the dam will fail within the first 50 years.

Alternatively and more usefully, if P_r is the risk that a designer is willing to accept that X will occur within r years, Eq. (6.23) may be solved for a design value T_D of the return period. The results are given in Table XIX for various values of P_r and r (195). Thus, for example, if

TABLE XIX

REQUIRED RETURN PERIOD T_D FOR RISK P_r THAT AN EVENT WILL OCCUR
AT LEAST ONCE IN A PERIOD OF r YEARS[a]

Risk P_r (%)	Return period r					
	2	5	10	20	50	100
75	—	4.0	6.7	15	36	73
50	3.4	7.7	15	29	73	145
25	7.5	18	35	70	174	348
15	13	31	62	124	308	616
10	20	48	95	190	475	950
5	40	98	196	390	976	1949
2	100	248	496	990	2475	4950
1	198	498	996	1992	4975	9953

[a] Kendall (195).

the assigned risk is 0.05 that the event will not occur in the next 50 years, the corresponding return period is 976 years.

For $r \gtrsim 10$ and $P_r \gtrsim 0.5$, the solution of Eq. (6.24) is approximated by Eq. (6.25),

$$(6.25) \qquad\qquad T_{\mathrm{D}} = r[1/P_T - \tfrac{1}{2}]$$

Equation (6.25) yields a design value T_{D}. This must be smaller than the estimate of T obtained from the observed data (for the same X). The quantity T may be estimated empirically but the extent of historical records is often much shorter than the desired return period. For a variable with an observed return period of 10 years, it can be shown that the true return period lies between 50 and 3.7 years with 95% confidence. This is too wide a range. A more efficient method is to fit a theoretical frequency distribution to the subset; the estimate is then based on all members of the sample rather than on the extreme few.

If $F(X)$ is the cumulative probability of X, a normalized variable is defined, $y = (X - \alpha)/\beta$. The equation of best fit is then determined empirically, the most common form being the double-exponential or Fisher–Tippett Type I.

$$(6.26) \qquad\qquad F(X) = \exp(-e^{\pm y})$$

The negative sign applies for maxima, e.g., floods, and the positive sign for minima, e.g., droughts. Gumbel (196) has shown that for this distribution,

$$(6.27) \qquad\qquad \beta = \sigma_n/s_x, \qquad \alpha = \bar{X} - \bar{y}_n\beta$$

where $\bar{X}$ and s_x are the mean and standard deviation of X, and $\bar{y}_n$, σ_n are normalized functions that depend only on n (published in tables or in graphical form).

Gringorten (197) states that Eq. (6.26) "has worked even better than theory suggests." The distribution has been applied widely by Gumbel (196, 198), who developed a probability paper on which solutions of Eq. (6.26) were represented by straight lines. Alternatively, a tabular solution is available using the following equation:

$$(6.28) \qquad\qquad X_T = \bar{X} + K(T, n) \cdot s_x$$

where

$\qquad X_T = $ the value of X with a given return period,

$\qquad \bar{X}, s_x = $ the mean and standard deviation of the observed series of extremes,

$\qquad K(T, n) = $ a tabulated function of return period and sample size.

Table XX (195) gives selected values of K. For example, suppose

TABLE XX

VALUES OF THE FUNCTION $K(T, n)$ FOR GIVEN VALUES OF T AND n[a]

Sample size n	Return period T									
	5	10	15	20	25	30	50	60	75	100
15	0.967	1.703	2.117	2.410	2.632	2.823	3.321	3.501	3.721	4.005
20	0.919	1.625	2.023	2.302	2.517	2.690	3.179	3.352	3.563	3.836
25	0.888	1.575	1.963	2.235	2.444	2.614	3.088	3.257	3.463	3.729
75	0.792	1.423	1.780	2.029	2.220	2.377	2.812	2.967	3.155	3.400
100	0.779	1.401	1.752	1.998	2.187	2.341	2.770	2.922	3.109	3.349

[a] Kendall (195).

TABLE XXI
VALUES OF $F(T, n)$ FOR GIVEN RETURN PERIOD T AND
SAMPLE SIZE n^a

n \ T	10	20	25	30	50	75	100
15	2.476	3.233	3.409	3.604	4.113	4.525	4.818
20	2.400	3.075	3.292	3.468	3.968	4.362	4.643
25	2.350	3.007	3.218	3.391	3.874	4.259	4.533
30	2.317	2.960	3.166	3.336	3.811	4.187	4.455
40	2.272	2.898	3.099	3.264	3.725	4.093	4.353
50	2.244	2.857	3.056	3.217	3.671	4.031	4.288
60	2.224	2.830	3.025	3.185	3.633	3.989	4.242
75	2.201	2.800	2.976	3.150	3.592	3.943	4.194
100	2.181	2.769	2.959	3.114	3.549	3.896	4.142

^{a} Kendall (195).

$n = 25$ years, $\bar{X} = 30°C$, and $s_x = 5°C$. Then the required value of X_T for a 100-year return period is $30 + (3.729) \times 5 = 49°C$.

For a given T, the quantity X_T is subject to sampling variations, and confidence intervals must be assigned. Assuming that the extremes do indeed belong to a Fisher–Tippett Type I distribution, Kaczmarek (199) has shown that for sufficiently large n, X_T is normally distributed with standard deviation $\sigma(X_T)$ given by

$$(6.29) \qquad \sigma(X_T) = F(T, n) \cdot s_x / \sqrt{n}$$

where $F(T, n)$ is another tabulated function, given in Table XXI (195). For example, if $n = 25$ years, $s_x = 5°C$, $T = 75$ years, then $\sigma(X_T) = 4.259 \times 5/\sqrt{25} \simeq 4.3°C$ and the 95% confidence limits are $(X_T = 8.6)°C$. If the return period T is only 10 years, $\sigma(X_T) = 2.35 \times 5/\sqrt{25} = 2.4°C$ and the 95% confidence interval changes to $(X_T \pm 4.8)°C$.

The Fisher–Tippett Type II equation is given by the equation

$$(6.30) \qquad F(X) = \exp[-(x/\beta)^{-\gamma}]$$

where γ is an empirical constant. Thom (200) has used this to describe the peak annual gust u'. The scale parameter β was found empirically to be related to the highest mean monthly wind speed by the equation

$$6.31) \qquad \beta = (347.5\bar{u} + 364.5)^{1/2} - 19.1$$

The shape parameter γ had two values, one of 9.0 for extreme wind distributions for extratropical storms and thunderstorms, and one of 4.5 for tropical storms. For locations in the United States subject to hurricanes as well as extratropical storms, Thom combined the two forms as follows:

$$(6.32) \qquad F_{ET}(u') = (1 - P_T)F_E(u') + P_TF_T(u')$$

where subscripts E, T refer to extratropical and tropical storms and P_T is the probability of an annual extreme gust from a tropical storm.

The usefulness of a design peak wind has been questioned because the historical data are dependent on the response characteristics of the anemometer and on the local exposure. In addition, the resonance of buildings is frequently of more concern than the static loading.

Other applications of extreme-value theory include snow-load design (a log-normal distribution), intense rainfall peaks (Poisson distribution), air conditioning extremes (Type-I maxima), and winter heating design (Type-I minima).

The importance of visual inspection of the plotted cumulative frequencies must be emphasized. Tables XX and XXI cannot be used, for example, unless there is evidence that the extremes do in fact approximate a Type I distribution. Commenting on series that consist of annual maximum discharges from rivers, Gumbel (198) has said that the best-fitting distribution may not be the same for one river as for another, a fact that has not yet been explained. In addition, physical constraints sometimes place an upper limit on the extremes, as illustrated in Fig. 51, which represents the distribution of January maximum temperatures in Washington, D. C. (197). Landsberg (personal communication) believes

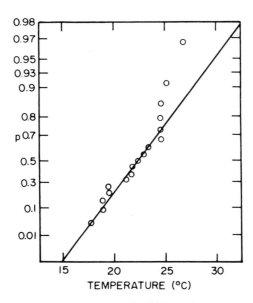

Fig. 51. The frequency distribution of the highest January temperature in each consecutive 5-year period in January 1875–1960 at Washington, D. C. (197).

that most of these events occurred with brisk southerly winds and that a "ceiling" was set by properties of the source region together with strong vertical instability. Similarly, Tunnell (201) found a marked change in slope of the cumulative distributions of both wet- and dry-bulb 0900 LST temperatures in Aden, which he has attributed to sea breezes or subsidence.

Finally, Benson (202) has shown that the average probability of all events that equal or exceed the k-year event is $\frac{1}{2}k$. This result is independent of the assumed form of the extreme-value distribution.

6.7. PEAK-TO-MEAN RATIOS

A widely adopted statistic in air pollution studies is the peak-to-mean ratio P/M, which is the peak concentration P averaged over some short interval of time t_P divided by the mean M for some longer period t_M. The ratio depends among other things on the values of t_P and t_M. For $t_P = t_M$, the ratio is one. The main interest, however, is usually in very small values of t_P.

A classical problem is the diffusion of pollution from a point source in a uniform flow. A number of models predict the resulting ground-level concentrations averaged over an hour or so (16). As the sampling time is reduced, however, the variation of concentration (or dosage) becomes uneven, and excursions from the mean may become large. An example is given in Fig. 52 (203), which shows the cross-wind distribution of

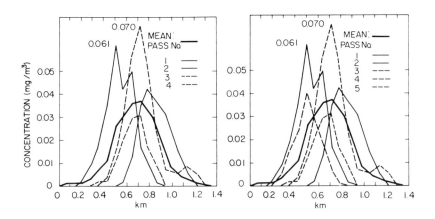

Fig. 52. Actual cross-wind distributions of concentrations from an elevated source at Brookhaven, New York; source height was 105 m and center-line down-wind distance was 2.5 km; the distribution found in pass no. 5 (right-hand panel) could not be predicted from the data of passes 1–4 (203).

concentrations. The instantaneous peak values are not predicted by the standard models, although they are of considerable biometeorological importance. A physical approach has been developed by Gifford (204, 205), who separates diffusion from a continuous point source into two components, one due to meandering of the entire plume about its mean position, the other due to diffusion away from the instantaneous center-line position by smaller-scale turbulent fluctuations. Both Gifford's model and Singer's experimental data (203, 206) indicate that for an averaging time of about an hour for t_M and for a steady wind direction:

(a) The peak-to-mean ratio approaches unity as distance from the source increases.

(b) The ratio increases with increasing angular displacement from the center-line.

(c) The ratio is larger for an elevated than for a ground-level source.

(d) The ratio is larger in unstable than in neutral temperature strati-fication (little information is available concerning the inversion case).

(e) The ratio is large (as much as 200) in the zone of aerodynamic downwash behind a building and near a chimney. For long travel dis-tances, however, peak-to-mean ratios are smaller over rough than over smooth terrain, because of more effective mixing by the turbulent wind.

Singer *et al.* (206) have noted that observed peak-to-mean ratios at a point may be fitted empirically by the equation

$$(6.33) \qquad P/M = (t_P/t_M)^{-\alpha}$$

where α depends on downwind distance, crosswind displacement, height of emission, and atmospheric stability. Subsequently, Hinds (207, p. 11) has justified this relation theoretically. Equation (6.33) is reminiscent of an empirical formula suggested by Mitsuta (208) for determining the peak wind gust to be used in the design of structures:

$$(6.34) \qquad P_t/M_{300} = (t/300)^{-\alpha}$$

where P_t is the peak t-sec gust, M_{300} is the mean wind averaged over 300 sec, and α is an empirical constant ($t < 300$).

More recently, Csanady (209) has emphasized that a peak can only be specified usefully in terms of its probability of recurrence. The pub-lished experimental values of P/M are based on very limited data; there is no doubt that if the diffusion trials had been replicated many times, larger values of P/M would have been reported. Csanady has developed a model that predicts that at the center of the plume, the mean concen-tration $\bar{C}$ averaged over the time period t_M is given by

$$(6.35) \qquad \bar{C} \simeq \langle \overline{C^2} \rangle^{1/2}$$

where $\langle \overline{C^2} \rangle^{1/2}$ is the root-mean-square value of the concentration C obtained from consecutive measurements each averaged over short time intervals t_P. Csanady suggests further that for a given center-line position,

$$(6.36) \qquad\qquad \phi = 1 - \exp(-C_0/\bar{C})$$

where ϕ is the probability that $C \gtrless C_0$, C_0 being any specified value of C.

For the outer edges of the plume, meandering causes the concentration to be zero occasionally. An intermittency factor q is therefore introduced, where $(1 - q)$ is the probability of zero concentration. Csanady then speculates that

$$(6.37) \qquad\qquad \phi = 1 - q \exp(-qC_0/\bar{C})$$

where $q \to 0$ for large angular displacements. Order-of-magnitude estimates of q are 0.5 and 0.1 at lateral distances of 1.5 and 2.5 standard deviations away from the plume centerline, respectively.

These models apply to stationary conditions over an hour or so and have not yet been used to study the multiple-source problem of a city. However, there have been many investigations of urban frequency distributions of peak-to-mean ratios for wide ranges of t_P and t_M; an example is given in Fig. 53 (210). The abscissa is the averaging time t_P

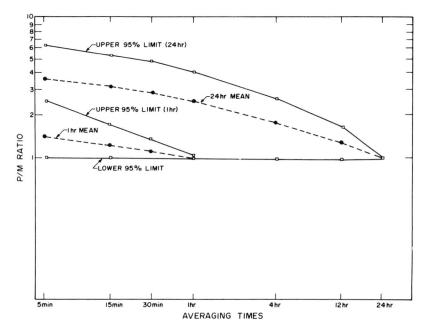

Fig. 53. Peak-to-mean ratios for an urban pollution sampling station for various averaging times for the peaks (5 min to 24 hr) and for the means (1 and 24 hr) (210).

for the peaks; the ordinate is the P/M ratio. Included within the figure are data for t_M of 1 hr and 24 hr. The dashed lines are the average values of P/M while the solid lines enclose areas within which 95% of the values fall.

Another kind of peak-to-mean ratio may be important in urban areas, viz., the highest of the point concentrations divided by the areal mean. This ratio has not yet been studied.

The P/M ratios for $t_M = 24$ hr in Fig. 53 are conceptually different from those discussed by Gifford, Singer, and Csanady. Nevertheless, there is some practical interest in large t_M, including values so large that the quantity M approaches its climatological average.

Barry (211) has examined the most general form of the equation for diffusion from a point source:

$$(6.38) \qquad D/Q = Wt$$

where D is dosage, Q is source strength, t is time, and W is a complicated function of "weather" and position (W is in units of seconds per cubic meter). For a sufficiently long period (years) to ensure climatological stability, W approaches a constant value W_∞ at any given site. Thus when a maximum permissible dosage MPD is specified by health authorities,

$$(6.39) \qquad MPD/Q = W_\infty t$$

This equation may be used to calculate the degree of containment required for an effluent and, if necessary, the size of the surrounding exclusion area, given values of W_∞. Barry believes that an effort should be made to determine W_∞ in various meso- and macroclimatic settings. He, for example, has estimated W_∞ at four sampling locations in the vicinity of the nuclear reactor at Chalk River, Canada, where Q is known. The quantity W_∞ ranges from 5.9×10^{-7} to 0.5×10^{-7} sec/m^3 at distances of 0.6–4.8 km from the chimney. Because engineering design requires only order-of-magnitude estimates, this range of values is acceptable.

Of equal interest is the question of protection in case of accidental releases. There is no way of knowing in advance the associated value of W, i.e., of knowing whether an accident will occur with east or west winds, in winter or summer, and in the daytime or nighttime. Barry therefore urges that the cumulative frequency distribution of W be obtained for a relevant averaging time. The distribution may be used in conjunction with the following equation:

$$(6.40) \qquad P_c = P_a \cdot P_{W'}$$

where

P_c = the probability of unacceptable consequences of the release (determined by health authorities),

P_a = the probability of an accident (estimated by the engineers),

$P_{W'}$ = the probability of W exceeding some given value W'.

Given P_c and P_a, Eq. (6.40) may be solved for $P_{W'}$. The value of W' is then found from the frequency distribution of W and is substituted into Eq. (6.38) to yield Q, given the maximum permissible dosage D for the time interval of concern.

Although P_c is very small, so also is P_a. Furthermore, because $P_{W'} < 1$, the product $P_a \cdot P_{W'}$ may often be sufficiently small that extremely small values of $P_{W'}$ are not required. If this be the case, the uncertainty in determining the shape of the "tail" of the W distribution will not arise. Barry's approach applies not only to nuclear reactors but also to industrial plants that must protect against the possibility of ruptures of fluoride or H_2S containers, for example, and complements the more conventional methods of dealing with this kind of problem.

7/STATISTICAL METHODS II: TIME SERIES

7.1. NORMALS

Occasionally a geophysical time series begins or ends abruptly, as when a river is dammed to form a lake. More frequently, the series is almost infinite in length, although only a small segment is available for study. The investigator has no data for the future, and his record of the past is truncated.

A *stationary* time series is one having constant moments, including a constant mean and variance. This condition rarely occurs in geophysics. Indeed, Lamb (212) believes that the term "climatological normal" is inappropriate and should be replaced by "average value for a datum period." The selected segment of record must be specified in terms of its length and in relation to a reference time. It is not meaningful to publish 30-year averages without indicating which years are included.

As the length of record increases, the statistics usually become more stable, although the stability obscures trend. A 30-year mean temperature at an observing point in the city of Detroit represents the temperature for that particular period and no other. The mean is useful for spatial comparisons with other locations having concurrent data. Because of increasing urbanization, however, the latest 10 years of observations may provide better forecasts of future temperatures than are given by the longer record; this consideration must be balanced carefully against the possibility that the 10-year record is too short to eliminate the "natural" variability of the weather, although the word "natural" is not easy to define in this context.

These considerations apply to time scales of minutes as well as to those of centuries. For purposes of illustration, however, a study by Craddock and Grimmer (213) is instructive. Given a sequence of mean annual temperatures $T_1, T_2, \ldots, T_{n-1}$ at a station, the problem is to find the most likely value of T_n.* The least-squares best estimate T_n' is given by

$$(7.1) \qquad T_n' = (1/k)(T_{n-1} + \cdots + T_{n-k})$$

where k is an integer to be determined by minimizing the variance s^2:

$$(7.2) \qquad s^2 = [1/(n - k)] \sum_{i=k+1}^{n} (T_i - T_i')^2$$

The historical records for 79 stations in the Northern Hemisphere were examined in this way. Craddock and Grimmer found that the value of k was variable but in general was smaller for locations with inhomogeneous records, for which averaging times of 10–15 years yielded minimum values of s^2. Court (214) has reviewed critically a number of papers on this subject, and he recommends a 20-year averaging time for mean annual temperature. However, he suggests that in some engineering applications, there is need to seek the best estimator not just for 1 year in advance but for the lifetime of a proposed structure (30 years or more).

In other cases, the emphasis is on the selection of an averaging period that yields a stable mean and frequency distribution, rather than on the question of extrapolation. Some estimates given by Landsberg and Jacobs (215, p. 979) are shown in Table XXII, which indicates a geographic control of variability.

Spatial comparisons often require adjustment to a common reference period, say, the years 1931–1960. A few stations in a network may have incomplete data or may have weather records only for some earlier period, 1903–1917, perhaps. Recommended methods of adjustment have been described by Thom (216, pp. 7–11) in which overlapping observations from two or more stations in the same meso- and macroclimatic regimes are compared.

Let X_1, Y_1 be the means for the two stations for the period of overlap, and let X_0 be the known mean for the reference period; it is then required to estimate Y_0. For most meteorological elements, the difference method is used, i.e.,

$$(7.3) \qquad Y_0 = X_0 + (Y_1 - X_1)$$

In the case of precipitation (except in arid regions), the ratio method provides better estimates (an empirical finding):

$$(7.4) \qquad Y_0 = X_0(Y_1/X_1)$$

---

* A problem that is conceptually similar is the prediction of the one-minute wind for rocket launches, given values over previous 1-min intervals.

TABLE XXII
APPROXIMATE NUMBER OF YEARS TO OBTAIN A
STABLE FREQUENCY DISTRIBUTION[a]

Climatic element	Islands	Shores	Plains	Mountains
	Extratropical regions			
Temperature	10	15	15	25
Humidity	3	6	5	10
Cloudiness	4	4	8	12
Visibility	5	5	5	8
Precipitation amounts	25	30	40	50
	Tropical regions			
Temperature	5	8	10	15
Humidity	1	2	3	6
Cloudiness	2	3	4	6
Visibility	3	3	4	6
Precipitation amounts	30	40	40	50

[a] Landsberg and Jacobs (215, p. 979).

Thom has illustrated the procedure with examples from Geneva and Lausanne, in which adjusted means compare favorably with actual values.

7.2. VARIANCE AND SPECTRA

The variance of a time series may be used as a measure of variability. For a 30-year temperature record, this statistic can be calculated from 30 annual mean values, 30×12 monthly means, 30×365 daily mean values, or $30 \times 365 \times 24$ hourly means. Four different variances are obtained, the one determined from hourly values being the largest. This illustrates one fundamental consideration in the study of time series: the variability is a function of smoothing time.

More generally, let X be a continuous variable having smoothed values $X_1, X_2, \ldots, X_k$ over consecutive intervals of time, each interval of constant width w, called the *smoothing time*. The total length of the record is therefore kw. The statistics are denoted by

$\bar{X}$ = mean value of X,

X' = deviation of any X from $\bar{X}$,

$\overline{X'^2}$ = variance of X.

For illustrative purposes, consider the effect of doubling the smoothing time from w to $2w$. Because of the additional smoothing, the variance is reduced. The largest possible reduction is achieved when X_i are drawn from a table of random numbers; a lesser reduction occurs when there is a correlation between X_i and X_{i+1}. In the limiting case of repetition of each X_i (the series $X_1, X_1, X_2, X_2, \ldots, X_k, X_k$), an increase from w to $2w$ has no effect on the variance. Most geophysical time series include persistence; for example, the correlation between consecutive annual flows of the St. Lawrence River is 0.7 (217). For increasing smoothing time, therefore, the variance decreases more slowly than it would for a set of random numbers. This is the second fundamental consideration in the study of time series.

An appreciation of these two relatively simple yet important ideas assists in understanding the approaches that have been taken in time series analysis. On the one hand, a *Markov chain* has been used to model the correlation that exists between consecutive members of a set (see Section 7.8). On the other hand, a *spectrum analysis* is employed to partition the variance according to the smoothing time, a topic discussed below and in Sections 7.3–7.6.

Suppose the initial data consist of 30 years of hourly mean temperatures. Then variances can be computed conveniently for smoothing times w of 1 year, 1 month, 1 day, and 1 hr. Let the four variances be represented by Y^2, M^2, D^2, H^2. As w becomes smaller, the variance becomes larger because more and more of the fluctuations are retained, i.e., $Y^2 < M^2 < D^2 < H^2$.

The statistics may be displayed graphically in the form of a spectrum. The abscissa is *period* T and the ordinate is an estimate of *spectral density* $F(T)$ defined as follows

(7.5)
$$M^2 - Y^2 = \int_{1\,\text{mo}}^{1\,\text{yr}} F(T)\, dT, \qquad D^2 - M^2 = \int_{1\,\text{day}}^{1\,\text{mo}} F(T)\, dT,$$
$$H^2 - D^2 = \int_{1\,\text{hr}}^{1\,\text{day}} F(T)\, dT$$

The quantity $F(T)$ for a stationary random process is actually a continuous function, and its form can be revealed in more detail through finite differencing, i.e., by using a large number of values of w, such as 6 months, 3 months, 1 week, 3 days, and so on.

Because of the wide range of times contributing to atmospheric variability (seconds to centuries), the abscissa is often converted to a logarithmic scale. Equal areas may be preserved by noting that

$$\int_{T_1}^{T_2} F(T)\, dT = \int_{T_1}^{T_2} TF(T)\, d\ln T$$

Hence, the ordinate scale is $TF(T)$ if the abscissa is $\ln T$. An example is given in Fig. 54.

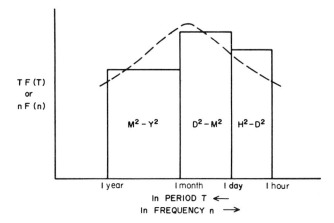

Fig. 54. Variance density spectrum of a time series in terms of period T and of frequency n.

Sometimes the area under the curve is normalized to unity by dividing by the total variance. This permits a relative comparison of several time series. It is to be noted, however, that the curve in Fig. 54 terminates on both sides of the diagram because (to the left) the segment of record is not infinitely large and because (to the right) the smoothing time w is never infinitely small. If two series with different values of w and/or k are normalized, therefore, they should not be compared.

A period T (units of time) is the inverse of a frequency n (cycles per unit time). Thus, the abscissa in Fig. 54 may be scaled according to increasing frequencies rather than to decreasing periods. A 24-hr periodicity is equivalent to a frequency of 1 cycle per day. The curve in Fig. 54 has been called traditionally a *power spectrum* because the method was used initially to examine the noise in communication circuits. However, *variance density spectrum* is a more appropriate term.

7.3. THE BLACKMAN–TUKEY METHOD OF SPECTRAL ANALYSIS

Although a spectrum can be estimated by determining variances for a number of values of w, this method is rarely used, and has been described in Section 7.2 only to provide an interpretive introduction. Instead, the *autocorrelation* for successive lags is calculated; this is the correlation between X_i and X_{i+j}, where j is the lag number ($j = 0, 1, 2, 3, \ldots$). For $j = 0$, the correlation is unity, and if X_i are drawn from a table of random numbers, all coefficients of correlation for nonzero lag are zero. Time series containing persistence but not trend have autocorrelations that become negligibly small for a sufficiently large lag time (see

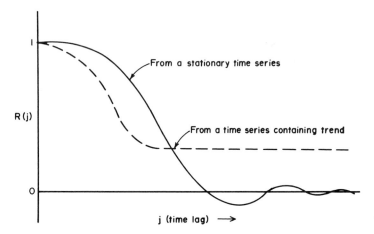

FIG. 55. Autocorrelation function for a stationary time series (———) and for a
time series containing trend (– – –).

Fig. 55). Consider, however, a series of annual temperatures from a
station that has become increasingly urban. A later value in the set is
higher on the average than an earlier one, resulting in a nonzero autocorrelation. In Fig. 55, the curve that levels off parallel to but above the
abscissa is typical of a time series containing trend. This occurs also
when the record includes only part of a low-frequency oscillation.

Formally, the autocorrelation function $R(j)$ is given by

$$(7.6) \qquad R(j) = \overline{X_i X_{i+j}} / \overline{X'^2}$$

The numerator of the fraction on the right side of the equation is called
the *autocovariance*. The *scale L* is defined as,

$$(7.7) \qquad \int_0^\infty R(j)\, dj = L \qquad \text{(units of time)}$$

Thus, the scale is finite only for a stationary time series. The quantity
L is the area under the $R(j)$ curve in Fig. 55. It can be shown (16, p. 4)
that the correlation function $R(j)$ is the Fourier transform of the
spectral function $F(n)$:

$$(7.8) \qquad F(n) = 4 \int_0^\infty R(j) \cos 2\pi n j\, dj$$

Hence, the spectral function may be determined from $R(j)$, using Eq.
(7.8).

To summarize the Blackman–Tukey method of estimating spectra,
the autocorrelation of the time series is calculated and then a Fourier
transform is performed. Both operations require access to a high-speed

digital computer. The procedure was developed by Blackman and Tukey (218) and has been described by Panofsky and Brier (45), Muller (219), and Mitchell (17).

7.4. COMPUTATIONAL CONSIDERATIONS

If 100 years of data from a stationary time series are available for analysis, the smallest frequency that can be examined profitably is one cycle per 20 years. The estimate of spectral density at this frequency is based on a sample of five 20-year segments of record. For one cycle per 50 years, on the other hand, the sample contains essentially only two data points, and thus is subject to large random variability. In general, the practical low-frequency limit is one cycle per $k/5$, although some investigators prefer a value of one cycle per $k/10$, where k is the number of observations.

At the high-frequency end of the spectrum on the right side of Fig. 54, the limit is called the *Nyquist* or *folding frequency;* this is $1/2w$, one fluctuation in twice the smallest data interval. For hourly observations of temperature, for example, the highest frequency that can be analyzed is one cycle per 2 hr.

By analogy with the spectrum of visible light, noise in communication circuits is classified as white, red, and blue. There is *white noise* when all frequencies contribute equally to the variance, e.g., when all X_i are chosen from a table of random numbers. *Red noise* occurs when low-frequency energy predominates, e.g., when the sensors are subject to calibration drift or to slow oscillations. *Blue noise* is associated with high frequency fluctuations, due perhaps to an unstable amplifier that produces random uncorrelated fluctuations from instant to instant. The three types are illustrated in Fig. 56.

Environmental time series frequently contain both red and blue noise. The latter is due to random errors of measurement, uncorrelated in time; the former is caused by low-frequency oscillations or trend. For a finite length of record, there is no way of distinguishing between a low-frequency wave and trend.

Aliasing (see Fig. 6) is a serious problem in spectral analysis. An individual X_i must not be an instantaneous value but must be pre-smoothed over an interval of time w, i.e., the time series must be modified by some smoothing technique such as running-mean averages. Otherwise, the high-frequency energy is shifted or "folded" into lower frequencies. An example has been given by Sabinin (220). His data consisted of temperature fluctuations in the thermocline of the tropical

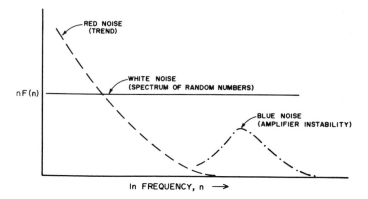

FIG. 56. Spectra of red, white, and blue noise.

Atlantic Ocean, measured with a thermograph suspended from an anchored buoy. Observations were taken at 5-min intervals over a period of 4 days. The results are shown in Fig. 57. The curve at the bottom of the diagram is the spectrum calculated from all the data. The two dotted curves are examples of spectra obtained from 5-min observations equally spaced once every 2 hr, i.e., they are based on only 1/24th of the data. It is possible to calculate 24 such spectra, the mean of which is

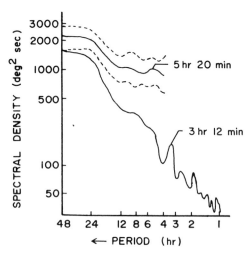

FIG. 57. Example of aliasing. The source data are consecutive 5-min values of temperature in the thermocline of the tropical Atlantic Ocean, measured over a 4-day period. The lower solid line is the spectrum obtained from all the data. The two dotted lines are examples of spectra obtained from 5-min observations equally spaced once every 2 hr. The upper solid curve is the spectrum obtained by averaging all the spectra of the type represented by the dotted curves (220).

given by the second curve from the top. Aliasing has introduced fictitious energy in all periods and has yielded a peak at 5 hr 20 min, which actually exists at 3 hr 12 min in the spectrum computed from all the data.

A spectrum analysis determines the energy in a series of frequency bands and yields a root-mean-square average amplitude for each band. No information is provided, however, concerning the distribution of amplitudes of individual fluctuations. For a given frequency band, the variance contribution may be the result of a few large fluctuations or many small ones.

A spectrum analysis fails also to detect phase shifts; this is both a weakness and a strength. If there is energy in a frequency band centered on 1 cycle per week, then a 7-day cycle that begins on Wednesday and another that begins on Sunday both contribute to this spectral density. As another example, a 3-day synoptic cycle may die out for a time but begin again out of phase with the earlier one; classical methods of harmonic analysis do not isolate such irregular rhythms.

Spectral estimates are themselves random variables. Even for a stationary time series, therefore, two segments of record cannot be expected to yield exactly similar variances. Thus, the "raw" estimates usually require smoothing, frequently by the "hanning" method in which,

$$P_0 = \tfrac{1}{2}(p_0 + p_1)$$

$$P_i = \tfrac{1}{4}(p_{i-1} + 2p_i + p_{i+1})$$

$$P_m = \tfrac{1}{2}(p_{m-1} + p_m)$$

where p_i is a "raw" and P_i a smoothed estimate, $i = 1, 2, \ldots, m$. It might be thought that smoothing could be achieved equally well by reducing the number of spectral estimates; "hanning," however, is preferable because "raw" estimates in adjacent frequency bands are essentially independent (in the absence of aliasing).

It may be shown (17, p. 40) that each spectral estimate P_i is distributed as chi square divided by the number of degrees of freedom df, with

(7.9) $$df \cong (2k - m/2)/m$$

where k is the number of observations and m is the number of spectral estimates. A 95% confidence interval can therefore be computed for each estimate. When there is a gap in the record, the autocorrelation can be calculated only for each unbroken segment, thus reducing the confidence in the low-frequency estimates. If there are only a few missing data, dummy values may be inserted (linear interpolation); by "a few" is meant possibly less than 5% of the total number of observations.

Sabinin (220) has considered the effect on aliasing of using a first-order sensor (defined in Section 2.6) with time constant λ. The signal is damped and it might be thought that this would have the same effect as presmoothing. However, if w is now the separation time between consecutive instantaneous observations:

(a) For $w = \lambda$, there is little aliasing but there may still be significant phase distortion.

(b) For $w > 3\lambda$, there is serious aliasing.

A recommended sampling procedure (219, p. 57) is to choose $w \leq \lambda/2$, i.e., for an instrument with a time constant of 20 min, observations should be taken every 10 min.

For consecutive observations smoothed over fixed intervals of time w, the effect of a sensor's time constant is negligible for $w \simeq \lambda/4$. However, when w and λ are of the same order of magnitude, an inverse filter such as that given by Eq. (2.15) should be applied to restore the signal. The theory has been given by Muller (219). If for some reason it is not convenient to do this and if the high frequencies are of interest, w should be chosen such that $w \simeq \lambda/4$.

Sometimes the initial data are space averages. Kaimal (221) has measured the temperature spectra obtained from two resistance thermometers, one 20 cm and the other 80 cm in length, both at a height of 9 m on a 2-m boom pointing into the wind on a tower. The spectral frequencies n were expressed in units of $n/\bar{u}$ (cycles/meter) where $\bar{u}$ is the mean wind; this is a common normalization practice that is supposed to remove the effect of wind speed. The ratio of the spectral densities was near unity for values of $n/\bar{u}$ less than 0.1, but the ratio decreased to $1/2$ for $n/\bar{u} = 1$, i.e., the 80-cm resistance thermometer attenuated much of the high-frequency energy.

Finally, data sets that are discrete rather than presmoothed values should be mentioned. For example, a series consisting of the annual dates of the freezing of a river has a Nyquist frequency of 1 cycle per 2 years: higher frequencies simply do not exist. An example is given in Fig. 58, a spectrum of the dates of Easter, based on "observations" for the years 1753–2000 AD, inclusive. This series does not contain "noise"; there is an exact although complicated way of determining the date of Easter in any particular year. The following features in Fig. 58 are of interest:

(a) Leap year results in a peak at 1 cycle/4 years but the variance density overlaps into adjacent frequency bands.

(b) The peak at about 1 cycle/2¾ years is puzzling at first glance. Examination of the time series, however, reveals a "saw-tooth" tendency

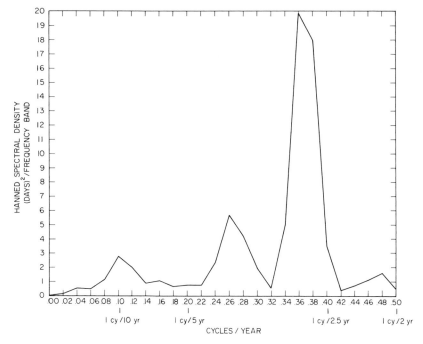

FIG. 58. Spectrum of the dates of Easter for the years 1753–2000 AD.

for alternating 2-year and 3-year periodicities, the net result being the large peak at intermediate frequencies.

(c) The 19-year lunar cycle should be present in Fig. 58 but spectral density resolution was poor at the lowest frequencies. A standard computer program with 25 lags was used.

There are many ecological examples of time series consisting of discrete data: the annual harvest, the annual bird count, and so forth.

7.5. OTHER METHODS OF ESTIMATING SPECTRA

There are other ways of obtaining spectra. One of these is the so-called *analog* method, in which the signal is processed through an analog computer equipped with band-pass filters. These filters attenuate frequencies that are both lower and higher than those in the "window" of interest. This operation may be performed in real time but because the calculations of spectral density estimates for different frequency bands are consecutive rather than concurrent, the variability among samples (including trend) may confound the analysis. Preferably, the magnetic tape containing the signal should be played back later.

The Blackman–Tukey method also sometimes employs band-pass filters but in this case the filters are mathematical in form. They are used to reduce the relative error of low-frequency spectral estimates; a discussion of this question has been given by Alekseev (222).

Of recent interest is the fast Fourier transform FFT, described by Cooley and Tukey (223). The method has three advantages. First, there is a significant saving in computing time: for a series of k values, the number of operations is of order $k \log_2 k$ as compared with k^2 for the conventional method of Section 7.3. Second, round-off errors are reduced. Third, more flexibility is possible in the selection of frequency bands.

In the FFT method, the time series is subdivided into c segments, possibly overlapping, each of length r, and the series becomes a matrix of c columns and r rows. The Fourier transform of each segment is then calculated and the results combined. As emphasized by Bingham *et al.* (224), "this 'factoring' saves arithmetic, and if repeated many times, saves much more arithmetic." The autocovariances may be computed, if desired, by a fast Fourier retransform, a much more rapid operation than direct calculation.

7.6. NONSTATIONARITY

The absence of trend is a necessary but not a sufficient condition for stationarity. Julian (225) notes that a time series may have a constant mean value but a variance decreasing with time, as in the case of a damped vibrating spring or pendulum given an initial displacement. As another example, consider a continuous time series of air temperature during a spell of clear weather. Superimposed on the strong diurnal cycle are microscale temperature fluctuations. This "noise" has a different quality in the daytime than at night, and it disappears almost completely during brief periods of transition from inversion to lapse shortly after sunrise, and vice versa shortly before sunset. Thus, a 1-week record of consecutive 1-sec temperature values would hardly constitute a stationary time series.

The effect of trend permeates the environmental sciences, invalidating many regressions when used for prediction from other data samples. This is particularly true if the predictors are interrelated or if they are derived quantities. Namias (226) discusses predictions of monthly mean temperatures, at 30 stations throughout the world in winter, based on hemispheric empirical orthogonal functions of 5-day mean pressures for the years 1899–1939. When the method was tested with independent data for the years 1948–1961, the predictions proved to be reasonably successful for a while but by 1957, all skill had vanished.

Trend may be removed by fitting a curve to the time series (linear or curvilinear regression) and subtracting or adding an appropriate amount from each value to obtain a new series. This device also is sometimes employed to eliminate a regular cycle that is overshadowing other neighboring cycles of interest. The more usual approach, however, is to choose band-pass filters that reduce the effect of frequencies both higher and lower than those being investigated. The selection of the most appropriate filter is not a simple decision, and the investigator should seek the assistance of a statistician. A useful discussion of this question has been given by Mitchell (17, pp. 48–57).

7.7. CROSS-SPECTRUM ANALYSIS

Tukey (227) suggests that when a mathematician is told that x causes y, he automatically thinks of the relation: $y = f(x)$. This is rarely true in the environmental sciences, however, because of lag. Instead

$$(7.10) \qquad\qquad y = f(x, \lambda)$$

where λ is the lag time.

One approach that has been tried is to introduce a lag time before beginning a statistical analysis. The best value of λ is obtained either from a physical understanding of the problem or by an empirical study using various lags. Brant and Hill (228), for example, found that weekly totals of hospital admission in Los Angeles for respiratory ailments were significantly correlated with 7-day total oxidant dosages 4 weeks earlier. Similarly, McCarroll (117) has examined the cross correlation of SO_2 levels in New York City with the incidence of eye irritation and coughs for different time lags (in days). He found that the highest correlation with eye irritation occurred with zero lag, indicating that the response was immediate. For coughing, on the other hand, the cross correlation with SO_2 concentrations was highest when a 2-day lag was used.

Useful as these studies are, there is a more formal and more meaningful way of examining the relation between two time-dependent variables. Given $X_1, X_2, \ldots, X_n$, and $Y_1, Y_2, \ldots, Y_n$, the covariance is defined as

$$\overline{X'Y'} = (1/n)\Sigma X_i' Y_i'$$

Then the partition of $\overline{X'Y'}$ by frequency bands is meaningful, i.e.,

$$\int_0^\infty \mathrm{Co}(n)\, dn = \overline{X'Y'}$$

where Co is defined as the *cospectrum*. As in the case of $F(n)$ in Section 7.2, this is sometimes normalized to unity by division by the product of the standard deviations of X and Y. In the frequency band dn, the area under the curve $\mathrm{Co}(n)$ is a measure of the contribution of that band to the total covariance.

There is always a possibility that two time series appear to be uncorrelated but in fact are highly correlated if phase lag is considered. Muller *et al.* (18), for example, have examined the relation between mean monthly lake levels at Harbour Beach, Michigan and monthly basin precipitation estimated from a large number of rain gauges. The period of record was from 1873 to 1963 inclusive. Figure 59 shows the phase lead (in degrees) of precipitation in relation to lake levels at each frequency. Apart from trend (very low frequencies) and the annual cycle, lake levels lag precipitation by approximately 100°, or one-quarter cycle. Thus, the time lag between lake level and precipitation is not constant but increases with the length of the oscillation.

Referring back to Section 2.6, Eq. (2.15) gives the phase lag ϕ for a first-order response system. As the frequency f increases, the quantity $(2\pi f\lambda)$ becomes very large and ϕ approaches a value of $\pi/2$. In Fig. 59, therefore, the fact that ϕ is approximately $\pi/2$ suggests that lake levels have a first-order response to basin precipitation.

In general, a cospectrum provides only partial information about the

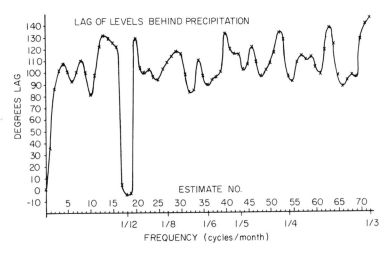

FIG. 59. Lag of monthly lake levels at Harbour Beach, Michigan, behind monthly basin precipitation (18).

relations existing between two time series. Additional insight is given by the *quadrature spectrum* $Q(n)$, in which each frequency of one variable is lagged by one-quarter period before commencing cross correlation with the other (45, p. 156).

Finally the *coherence* $CH(n)$ is defined:

$$CH(n) = \frac{Q^2(n) + Co^2(n)}{F_X(n)\, F_Y(n)}$$

The coherence varies from 0 to 1 and is an index of the strength of the relation between X and Y for frequency n.

7.8. PERSISTENCE

For a time series X_1, X_2, . . . , X_i with unit standard deviation, a *first-order linear Markov process* is defined by the equation

(7.11) $$X_i = \rho X_{i-1} + E_i$$

where ρ is a constant $(-1 < \rho < 1)$ and E_i is a random variable having a mean value of zero. Thus, each X_i depends on its own immediate past value plus a random component.

From Eq. (7.11) it follows that

(7.12) $$X_{i+1} = \rho X_i + E_{i+1}$$

Thus

(7.13) $$X_{i+1} = \rho^2 X_{i-1} + \rho E_i + E_{i+1}$$

and in general,

(7.14) $$X_{i+n} = \rho^{n+1} X_{i-1} + \text{other terms independent of } X$$

Because $-1 < \rho < 1$, the correlation between X_{i+n} and X_{i-1} approaches zero as n increases.

Gilman *et al.* (229) have computed the spectrum corresponding to this time series. Their results are given in normalized form in Fig. 60 for various values of ρ. The horizontal line represents the case $(\rho = 0)$ of a white spectrum; the other curves are instances of red noise. As a comparative example, Fig. 61 shows the spectrum of 12 years of daily precipitation amounts at Woodstock College, Maryland, after prewhitening to remove the annual cycle. Both the white and a best-fitting Markov spectrum $(\rho = 0.12$ by inspection) are given. Selected peaks are examined in Table XXIII, which shows that the significance of the peak near 2.8 days is increased when related to a Markov rather than to a white spectrum; the significance of the other peaks, however, is reduced. Gilman

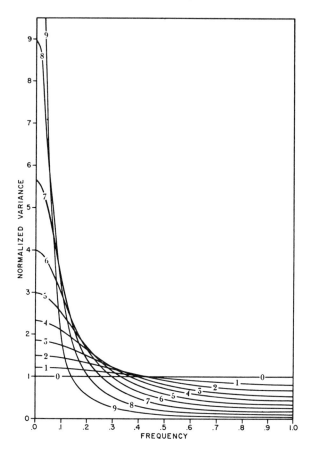

FIG. 60. Spectra of a first-order linear Markov process for various values of ρ in Eq. (7.11): one-lag autocorrelation, $9 = 0.90$, $8 = 0.80$, $7 = 0.70$, $6 = 0.60$, $5 = 0.50$, $4 = 0.40$, $3 = 0.30$, $2 = 0.20$, $1 = 0.10$, $0 = 0.00$. The ordinate is variance expressed as a fraction of the white noise value. The abscissa is frequency as a decimal fraction of maximum frequency, equal to half the data sampling frequency (229).

et al. comment that the relative importance of various peaks depends "on whether or not persistence in the series is taken into account, as it properly should be."

A Markov second-order process is given by the equation

$$(7.15) \qquad X_i = \rho_1 X_{i-1} + \rho_2 X_{i-2} + E_i$$

and so forth for higher-order relations. The limiting case of white noise is called a zero-order process.

Markov chains may be applied also in the analysis of dichotomous events. When there are only two possible outcomes (such as rain vs. no

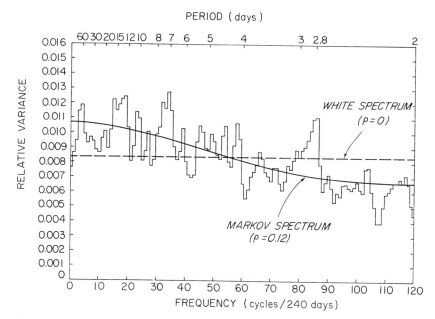

Fig. 61. Spectrum of daily total precipitation at Woodstock College, Maryland, 1943–1954: maximum lag is 120 days; ordinate is relative variance; area under curve is one. Population spectra for white noise and for Markovian red noise ($\rho = 0.12$) are included for comparison (229).

TABLE XXIII

STATISTICAL SIGNIFICANCE[a] OF SELECTED PEAKS IN THE SPECTRUM OF FIG. 61, ASSUMING ALTERNATIVE NULL HYPOTHESES[b]

Period in days (approx.)	Peak number of harmonics involved	Degrees of freedom	Null hypothesis	
			White noise ($\rho = 0$)	Red noise ($\rho = 0.12$)
2.8	1	73	0.95	0.99
7–8	3	219	0.999	0.99
12–17	4	292	0.999	0.95
53–69	1	73	0.99	<0.95

[a] Each harmonic of the spectrum has 73 degrees of freedom, df. The significance of a peak is determined by comparing the ratio of df times its amplitude to the amplitude of the white (or red) continuum at the same harmonic, with the percentage points of the chi-square distribution for the same value of df [see Eq. (7.9)].

[b] Gilman et al. (229).

rain), the chain is said to have *two states*, while a contingency table with k possible outcomes (no rain, plus $(k - 1)$ classes of total rainfall) may be modeled with a *k-state* chain.

As an example of the two-state chain, suppose that p_0 is the probability of a wet day, given the previous day dry, and $(1 - p_1)$ is the probability of a dry day, given the previous day wet. Then the probability of a dry spell of length m is

$$p_0(1 - p_0)^{m-1}$$

and of a wet spell of length n is

$$(1 - p_1)p_1^{n-1}$$

Gabriel and Neumann (230) have shown that this model fits observed sequences at Tel Aviv, although they emphasize that they have not established a physical basis for the relation.

Lowry and Guthrie (231) have examined two-state Markov chain models of precipitation, but they have extended the previous work to include orders 0, 1, 2, and 3. Their analysis suggests the following:

(a) With an increase in the threshold value separating wet from dry days, the order decreases.

(b) In many areas, first-order chains are satisfactory approximations; however, higher orders are more appropriate in regions which are "climatologically diverse within a season."

(c) Orders tend towards zero as precipitation days become rarer, i.e., the series approach a white-noise condition.

Lowry and Guthrie include a useful discussion of significance tests.

Gringorten (232, 233) has been developing a model for a sequence of hourly values that are normally distributed and that may be represented by a first-order Markov chain. The objective is to provide a very general framework for predicting the persistence of a condition above or below any arbitrarily chosen value. Sharon (234) notes that diurnal and annual cycles create difficulties but that the model is worth further development. In this connection, Godske (235) has undertaken a careful analysis of 3-hourly temperatures at Oslo and Bergen, and he concludes that these series are non-Markovian. As an example for Oslo in April, the 7 AM temperature is correlated more highly with the 9 AM temperature of the previous morning than with the 3 PM value, i.e., the correlation is larger with a lag of 22 than with a lag of 16 hr. This result was reproducible in every year that was examined.

In the studies described above, the Markov process has been treated as a statistical phenomenon. However, Eriksson (236) has recently suggested that a physical interpretation may be possible in some cases. Consider, for example, the following equation:

(7.16) $$dX/dt + aX = E(t)$$

where a is a constant and $E(t)$ is a random function of time not correlated with X. Equation (7.16) may be written in finite-difference form:

(7.17) $$(X_i - X_{i-1})/\Delta t + (X_i + X_{i-1}) \cdot a/2 = (E_i + E_{i-1})/2$$

Rearranging terms,

(7.18) $$X_i = [(2 - a\,\Delta t)/(2 + a\,\Delta t)]X_{i-1} + (\Delta t/2 + a\,\Delta t)(E_i + E_{i-1})$$

Equations (7.18) and (7.11) are similar. Thus, Eq. (7.16) represents a first-order Markov process. As an example, Eriksson shows that Eq. (7.16) may be used to describe the behavior of a ground-water reservoir: the variable X is the ground-water level, and the reservoir is assumed to drain at a rate proportional to $(X - X_o)$, plus a random component (X_o is a constant). In this case at least, a physical model has been developed that explains why the series X_i can be represented by a Markov process.

Returning now to the subject of log-normal distributions, suppose that successive members of a series obey the relation

(7.19) $$X_i - X_{i-1} = E_i X_{i-1}$$

This ressembles but is not the same as Eq. (7.11). Rewriting Eq. (7.19) and summing over n observations,

(7.20) $$\sum_1^n (X_i - X_{i-1})/X_{i-1} = \sum_1^n E_i$$

In integral form, the left side of Eq. (7.20) becomes

$$\int_1^n dX/X = \ln X_n - \ln X_1$$

Thus,

$$\ln X_n = \ln X_1 + \sum_1^n E_i$$

By the central-limit theorem, $\ln X$ is asymptotically normally distributed, and thus X is log-normally distributed. A more formal proof has been given by Aitchison and Brown (237), who also include many examples of log-normal time series that may be represented by Eq. (7.19).

8/PHYSICAL METHODS

8.1. DEDUCTIVE AND INDUCTIVE REASONING

Physical methods are based on deductive rather than inductive reasoning. Beginning with simple principles such as the conservation of matter and assuming precise sets of initial and boundary conditions, the present or future behavior of the system is predicted. For example, the evaporation rate from a lake is inferred from a balance of all sources and sinks of energy at the surface, or from a budget of all gains and losses of water in and out of the lake.

A physical model sometimes includes statistical quantities. In studies of diffusion of smoke in a uniform turbulent flow, for example, the motion of an individual particle is unpredictable; however, the time or space distributions of the positions of many particles may be specified statistically. The means and variances can be included in physical relationships with other factors such as wind and distance from the source.

A problem is often idealized in order that it can be specified in mathematical terms, and in order that the resulting equations can be solved. Some of the important variables frequently are kept constant in the simplified model; for example, a heat balance study may include the assumptions that solar radiation flux and the mean wind speed are constant.

Physical methods are important tools in geophysical processes. Their application to the life sciences, however, is not without difficulty. Although the heat balance or water budget of a living organism can be determined, the response of the cells to a particular environment is not specific. Because an individual reacts differently on separate occasions to the same heat loading or wind chill, and because the possibility of adaptation is frequently present in consecutive experiments, the physiologist rarely obtains reproducible results except when replications are averaged.

The alternative is to use inductive reasoning. A number of examples have been given in previous chapters, and the method serves a useful purpose. Consider, for instance, the study by Wayne *et al.* (238) of the performance of high-school cross-country runners in 21 competitive meets in Los Angeles from 1959 to 1964. Because of trend (running times tended to improve as the season progressed), team performance was expressed as the percentage of boys who failed to improve on their previous times. This quantity was high when oxidant levels 1 hr before each race were also high, and the correlation proved to be significant. The authors suggest that high oxidant values reduce performance levels because of physiological changes or because of decreased motivation resulting from discomfort. This line of reasoning appears frequently in scientific papers. Most people would accept intuitively the claim that pollution stress has an effect on athletic performance but the study does not provide conclusive proof and must be verified with detailed physiological measurements. When an inductive argument is associated with an intuitively plausible explanation, as in this example, the method is said to be *heuristic*, i.e., alternative explanations have not been examined and rejected in turn. Statistical methods (Chapters 6 and 7) are inductive; their power arises from the fact that they yield confidence limits for accepting or rejecting a hypothesis, provided that certain sampling criteria are met.

Inductive arguments *per se* are often useful. What must be criticized, however, is improper application, which can be illustrated with a trivial example. The fact that the sun has risen every day in recorded history suggests that there is every likelihood that the sun will continue to rise on future mornings. The prediction would, of course, fail if made by someone subsequently flown to the arctic in midwinter. Without an astronomical model, he might wonder whether the absence of daylight was a unique feature of his new location, correlated perhaps with the cold, or whether the sun had not risen also in his homeland. In this case, the conditions surrounding the experiment have been changed, and extrapolation is unsound.

8.2. DIMENSIONAL ANALYSIS

A physical problem can be simplified frequently by a dimensional analysis. In the hollow cylinder example of Section 6.4, suppose that the investigator is given the density D, the height H, and the inner and outer radii, R_I and R_O. Suppose further for the sake of illustration that the formula for the area of a circle is unknown. The following reasoning may be applied to simplify the problem of estimating the mass M. From the definition of density

$$M = DV = DHA$$

where V is the volume of the cylinder, A is the area of the end ring (H = height, D = density). Because the area in turn must be a function only of R_I and R_O,

$$M = DH{\cdot}F(R_I, R_O)$$

where $F(R_I, R_O)$ is an unknown function. The problem may be simplified further by introducing a new variable $y = M/DH$, with dimensions of (length)^2, i.e.,

$$y = F(R_I, R_O)$$

and the number of variables has been reduced from 5 to 3. Dimensional reasoning is invoked next to assert that $F(R_I, R_O)$ must have the same dimensions as y. This statement cannot be proved, although physical laws are always formulated in such a way that they are dimensionally homogeneous.

Having decided that $F(R_I, R_O)$ has dimensions of (length)^2, the simplest form may be assumed for an initial multivariate analysis:

$$y = a_1 R_I R_O + a_2(R_I)^2 + a_3(R_O)^2$$

There is, of course, no way for the analyst to know in advance whether the function contains additional terms with dimensions of (length)^2, e.g., $a_4 R_I R_O \exp(R_I/R_O)$, although a preliminary graphical analysis might help. The important points to be emphasized are that the correct five variables have been selected, that three of these, M, D, and V, have been combined into one, and that the dimensions of the unknown function $F(R_I, R_O)$ have been determined.

The dimensional method is widely used in geophysics. Although the predictions always require experimental verification, they nevertheless often lead to so-called *universal* relations. In contrast, statistical regressions need not be dimensionally homogeneous, and they provide little insight into the fundamental physical relations. In this and other biometeorological examples, therefore, an attempt should always be made to convert the variables to a dimensionally homogeneous form. The analysis proceeds as follows.

An environment containing no limiting factors is examined first, even though the ultimate interest may be in stress. The analysis is restricted to short periods of time with steady-state conditions in order to avoid the nonlinear interaction effect mentioned in connection with Eq. (2.1). The relevant variables for this ideal case are then listed and combined, wherever possible, as in the hollow cylinder problem $(y = M/DH)$. Despite such preliminary maneuvers, however, a solution may still be far from obvious. The investigator is therefore committed to a regression analysis. Even so, and this is a point generally overlooked, the resulting equation may often be interpreted in a physically meaningful way by assuming a dimensionally homogeneous form, i.e., by *assigning dimensions to the constants*. For example, suppose that leaf and air temperatures, T_L and T_A, and net radiation Q_N are found to be connected by a linear regression:

$$(8.1) \qquad T_L = aT_A + bQ_N + c$$

The quantity a is dimensionless and is perhaps therefore a universal constant (a speculation requiring experimental verification). The quantity b may be interpreted as having the dimensions °C cm^2 min/cal. It will be seen in Section 8.7 that this in fact matches the dimensions of the parameter $(r/\rho c_p)$, where r is resistance (minute per centimeter), ρ is air density (g/cm^3), and c_p is specific heat of air at constant pressure (calories per gram degree Celsius). The equation is therefore rewritten as follows (converting temperature to degrees Kelvin):

$$(8.2) \qquad T_L = a_1 T_A + b_1 Q_N (r/\rho c_p) + c_1$$

where a_1, b_1 are presumably universal constants, if temperature is in degrees Kelvin. The equation contains a third variable r, and in one sense the problem seems further from solution than before; nevertheless, a feature of the relation is suggested that may lead to a deeper insight into the problem. The behavior of the quantity r can be studied, and its variation with wind speed determined, for example.

For purposes of illustration, Eq. (8.1) has been chosen deliberately because it is in fact incomplete. Transpirational cooling Q_E is another controlling mechanism for leaf temperature, and Linacre (239) has derived a physical relation of the form

$$(8.3) \qquad T_L = T_A + (Q_N - Q_E)(r/\rho c_p)$$

Thus, the regression coefficients a_1 and b_1 are both unity. This illustrates an interesting "property" of dimensionally homogeneous equations; constants usually are of about the same order of magnitude and often have a value near unity, although no reason can be advanced. The investigator may view with suspicion a dimensionally homogeneous equation in which the constants do not obey this rule of thumb.

Returning now to Eq. (8.1), a little physical insight would have indicated qualitatively that leaf temperature must depend partially on transpirational cooling. Having said this, however, it must also be noted that Q_E is not an easy quantity to monitor, and that in the first instance, the data might be limited to measurements of T_L, T_A, and Q_N. The statistical regression, Eq. (8.1), could, however, be rewritten profitably as follows:

(8.4) $$T_L = aT_A + bQ_N + c + \epsilon$$

where ϵ is a residual, the departure of each individual T_L from its best-fit value, i.e.,

(8.5) $$\bar{\epsilon} = 0$$

The temperature should be in degrees Kelvin, of course. The quantity ϵ is a new variable whose behavior may be examined graphically as a function of other meteorological elements such as wind speed and vapor pressure deficit. It should be noted, however, that if the data contain relatively large random instrument errors, the relation becomes blurred. For example, if an unshielded thermocouple yields errors that are functions of wind speed and/or of Q_N, there is no way of disentangling instrument errors from the significant physical relations.

Although empiricism is sometimes unavoidable, even in the absence of stresses, the parameters of best fit often can be interpreted when they are assigned dimensions that make the equation homogeneous. Monteith (240), for example, has considered the well-established empirical relation

(8.6) $$p = (a + b/I)^{-1}$$

where p is the rate of photosynthesis (grams of carbohydrate per square meter of leaf area per hour) and I is light flux (ly/hr). Monteith has interpreted the parameters a, b in the following way. The quantity a has dimensions of m²hr/g or more meaningfully $(hr/m)/(g/m^3)$, i.e., it is a resistance multiplied by a volumetric index of crop density. The quantity b has dimensions of cal/g and is inversely proportional to the quantum efficiency of photosynthesis in very weak light. This interpretation, if correct, provides a rational basis for designing new experiments.

The analysis may now be extended to include growth-limiting factors. Sometimes the way to proceed is evident from equations such as Eqs. (8.2) and (8.3) by permitting, for example, the resistance to become very large. In other instances, a dimensionless factor is added: the width of stomatal opening as a fraction of the maximum possible, the ratio of actual leaf area to that of a healthy crop, or the ratio of respiration to photosynthesis (in the same units). This method is far more likely to have universal applicability than the determination of a large number of regression equations, one for each species, one for each climatic region, and so forth.

In micrometeorological studies, an index of vertical stability is the *stability ratio* SR (10, p. 82):

$$(8.7) \qquad SR = 10^5(T_2 - T_1)/\bar{u}^2$$

which has units of °C sec^2/cm^2. The quantities T_1 and T_2 are the temperatures at two levels and $\bar{u}$ is the mean wind speed at some intermediate level. This is a simplified form of the *Richardson number* Ri (nondimensional):

$$(8.8) \qquad Ri = \frac{g[(\partial T/\partial z) + \Gamma]}{T(\partial \bar{u}/\partial z)^2}$$

where Γ is the dry adiabatic lapse rate, g is the acceleration due to gravity, and T is temperature in degrees Kelvin. Equation (8.7) is often preferred to Eq. (8.8) because of the difficulty in estimating the vertical gradient of wind.

This example has been introduced for two reasons. In the first place, it is to be noted that Eq. (8.7) can be made nondimensional without difficulty:

$$(8.9) \qquad SR = \frac{gz^2(\partial T/\partial z)}{T\bar{u}^2}$$

Whether Eq. (8.9) is more useful than Eq. (8.7) is quite another question; the point to be stressed here is that a dimensional index can indeed be transformed to another dimensionless one, providing more generality.

Second, the Richardson number is an example of a large array of dimensionless quantities that have proved to be valuable. Others include the Prandtl, Froude, Schmidt, Grashof, and Reynolds numbers. When the appropriate ones are kept constant, the predictions become universal. For a given magnitude of the Reynolds number Re in a pipe, for example, a correct prediction of whether the flow is laminar or turbulent can be made, irrespective of the speed u, the diameter of the pipe D, and the viscosity of the fluid v (Re $= uD/v$).

Dimensional analysis is not a panacea for all biometeorological problems. However, there is often much to be gained and little to be lost by trying the method.

8.3. MODELING

Many physical processes which differ in scale nevertheless obey the same governing equations. For example, meteorological variations in a field of potatoes are quite similar to those in a tall forest; the methods for calculating fluxes are identical. This general principle is exploited when scale models are constructed to simulate the natural environment. In fact, a model of a model is sometimes built; wind tunnels are becoming so large that a "scale model" is often tested in advance of construction of the wind tunnel itself.

Geometric similitude occurs when the ratio of length scales between model and prototype is constant. Although an obvious initial objective, this is not always possible or desirable. In the case of a model of a river or of the atmosphere, for example, the fluid is spread so thinly that surface tension becomes important in the laboratory simulation. Models therefore are sometimes distorted, although they always preserve *homologous similitude;* for each point in space and time in the prototype, there is one and only one similar point in the model.

As an elementary example, suppose there is a need to determine the mass of a very large homogeneous rock of irregular shape. A small plasticine model is built with geometric similitude. Since the variables of importance are density D, volume V, and mass M, the dimensionless group that must remain constant is DV/M. Let $D' = k_D D$ and $L' = k_L L$ where D'/D, L'/L are the ratios of densities and lengths in the model and in the prototype. The quantities k_D and k_L are called the *scale factors.* If the mass of the model is M', then

$$M = M'/k_D k_L^3$$

On the other hand, if the irregular object is about a kilometer in length but only 100 m wide and 10 m high, geometric similitude is difficult to maintain; separate scaling factors k_x, k_y, and k_z are therefore introduced for length, width, and height, respectively. Thus,

$$M = M'/k_D k_x k_y k_z$$

The mass of the prototype can again be determined from the mass of the model.

Consider a moving particle, whose positions are given by distances x, x' at times t, t' in the prototype and the model, respectively, where $x' = k_x x$ and $t' = k_t t$. Then the velocity of the particle has a scale factor k_x/k_t, its acceleration k_x/k_t^2, and its nth derivative k_x/k_t^n. This is called *kinematic similitude,* but clearly, the scaling can quickly become uncontrollable if the motions are very complicated.

A more successful approach includes not only homologous and kinematic but also *dynamic similitude,* i.e., the ratios of forces between model and prototype are kept constant. Sometimes these forces are known. The transition from laminar to turbulent flow is dependent on the ratio of inertial to viscous forces, for example. In other cases, if the governing equations can be specified, a transformation to dimensionless form reveals the scaling factors immediately. Finally, it is often necessary or convenient to employ the *method of dimensions.* In the last case, the relevant variables are listed and then combined into a dimensionless set, either by inspection or by a more formal method (241). Dynamic similarity is achieved if the dimensionless groups have the same values in the model as they have in the prototype.

A typical application is the modeling of chimney downwash of a heated plume during strong winds. Lord and Leutheusser (242) have described the problem in some detail. Dimensional analysis leads readily to the following rather general functional relation:

(8.10) $$F_1(x/L, z/L, \delta/L, z_0/\delta, R, \text{Re}, \text{Ri}) = 0$$

where

x, z = any set of coordinates defining any point of the plume,
L = a characteristic length such as stack height,
δ = the thickness of the surface boundary layer,
z_0 = a measure of the surface roughness,
R = the ratio of gas exit velocity to wind velocity,
Re = the Reynolds number (nondimensional),
Ri = the Richardson number (nondimensional), an index of buoyancy forces.

There are thus seven independent dimensionless groupings; the relative importance of each and the precise form of the function F_1 can only be found by experiment.

The subsequent analysis described by Lord and Leutheusser may be summarized as follows:

(a) The Reynolds number is omitted, its value proving to be unimportant in strong turbulent flow.

(b) The ratios δ/L and z_0/δ are varied, and shown experimentally to have only a second-order effect (these quantities are highly significant in studies of wind loading on buildings).

(c) The Richardson number is difficult to model in most wind tunnels. Lord and Leutheusser suggest that it can be neglected, however. During strong winds, turbulent mixing is so vigorous that the atmospheric lapse rate is near the adiabatic and Ri is near zero.

Equation (8.10) therefore reduces to

(8.11) $F_2(x/L, z/L, R) = 0$

(d) There remains the question of selecting a scale factor. Lord and Leutheusser recommend

(8.12) $R' = R(T_A/T_G^{0.5})$

where R' is the ratio of gas velocity to wind velocity in the model, and T_A, T_G are absolute temperatures of the air and of the effluent gas, respectively. This is equivalent to assuming that the buoyant plume is replaced by another prototype plume, nonbuoyant but having the same initial momentum.

(e) Finally, the form of the unknown function F_2 in Eq. (8.11) must be determined experimentally.

8.4. WATER BUDGET

The conservation equations are widely used in the environmental sciences. The net change of water (or of energy, heat, or momentum) in a physical system is the difference between the gains and losses in unit time. A change in the amount of water in a lake, for example, is given by

(8.13) $N = P + I_r + I_g - E - O_r - O_g$

where

N = the change in the quantity of water in the lake (cm³ per day, month, or year),
P = the precipitation falling on the lake,
I_r = the inflow from rivers and runoff,
I_g = the inflow of ground water,
E = the evaporation from the lake,
O_r = the outflow through rivers,
O_g = the outflow to ground-water storage.

If all terms but one are measured experimentally, Eq. (8.13) may be used to estimate the unknown one. Because this is a small difference between relatively large values, the accuracy of estimate is not great.

In a similar fashion, the water budget of any volume can be formulated, including that of a human, an insect, a tree, or a watershed. Because of the difficulty in obtaining precise estimates of each term, the resulting equation is not always of predictive value. However, it does establish reasonable limits and ensures that an independent estimate of an individual term has the correct order of magnitude.

An illustrative example, the water budget of some field crops, has been given by Kharchenko and Kharchenko (243). Studies were made of wheat, corn, and fallow fields in the vicinity of the Dubovka Hydro-meteorological Observatory in the U.S.S.R., using Eq. (8.14) during periods when there was no percolation.

(8.14)
$$E = W_i - W_f + P - R - D$$
where,

E = evapotranspiration,
W_i, W_f = initial and final soil moisture averaged over the layer containing the plant roots,
P = precipitation,
R = runoff,
D = moisture exchange with the lower inactive soil below the root zone.

Soil moisture was determined to a depth of 0.5 m at four points and to 1.5 m at four other locations in each crop. Because of random variations of 20% in determinations of W_i and of W_f, Eq. (8.14) was believed to be useful for determining E only when the sampling interval was at least one month.

An independent estimate of evapotranspiration was obtained from small weighing evaporimeters, 1 m deep and 710 or 1000 cm² in area. Some comparative results are given in Table XXIV. The average discrepancy was 13%, within the limits of random errors in the estimation of the various terms in Eq. (8.14). In the case of the corn field, however, departures were as large as 107% because the evaporimeter frequently dried out to the crop wilting point. This kind of study is useful in judging the reliability of methods for determining irrigation requirements.

Dawson *et al.* (244) have estimated the water budgets of some small Australian lizards, maintained without water in a dry laboratory for periods of 70–96 hr. The budget is formulated in Eq. (8.15), all the terms representing losses:

(8.15)
$$N = -C + U + E_c + E_r$$
where

N = the net change in milliliters of water per 100 g of lizard per day,
C = consumption of water (zero in this case),
U = urinary water loss,
E_c = evaporation through the skin (cutaneous),*
E_r = respiratory evaporation.

---

* For an animal with sweat glands, this loss is sometimes subdivided into *insensible perspiration* (diffusion of water through the skin) and *sweat.*

TABLE XXIV

TOTAL EVAPORATION (MM) ESTIMATED FROM A WATER BUDGET OF THE UPPER METER LAYER OF SOIL AND FROM A SOIL EVAPORIMETER[a]

Year	Crop	Period of observation (from seedlings to harvest)	Total evaporation		Deviation $\frac{E - E_{ev}}{E}$ 100%	Evaporation rate during growing period (mm/day)
			By water balance E	By evaporimeter E_{ev}		
1955	Barley	16/IV–1/VII	243	235	3	3.1
	First-year alfalfa	10/V–29/VI	108	130	–20	2.6
	Spring wheat	18/IV–9/VII	138	160	–16	2.0
	Winter rye	22/III–29/VI	244	216	11	2.2
	Winter wheat	21/III–29/VI	213	218	–2	2.2
	Corn	10/V–30/VIII	245	298	–21	2.7
1958	Barley	25/IV–1/VII	294	268	9	4.0
	Spring wheat	26/IV–16/VII	314	302	4	3.8
	Winter wheat	7/III–10/VI	307	331	–11	3.5
	Corn	10/VI–20/VIII	271	214	21	5.2
1959	Barley	22/IV–28/VI	121	96	21	1.4
	Spring wheat	22/IV–8/VII	120	150	–25	3.2
	Winter wheat	16/IV–8/VII	159	140	19	1.7
	Corn	18/V–28/VIII	67	143	–107	2.0
1961	Winter wheat	20/III–28/VI	293	317	–8	3.2
	Winter wheat (collective farm field)	5/IV–28/VI	249	227	9	2.7
	Spring wheat	10/IV–6/VII	224	184	18	2.1
	Corn	15/V–7/VII	166	220	–32	4.1
1963	Barley	4/V–4/VII	156	169	–9	2.8
	Winter wheat	18/IV–8/VII	211	192	9	2.4
	Corn	8/V–17/VII	51	86	–70	1.2
	Buckwheat	20/V–17/VII	95	117	–20	2.0
	Peas	7/V–4/VII	103	129	–25	(5.4)

[a] Kharchenko and Kharchenko (243).

The quantity N was obtained by direct weight measurements, assuming that the weight gained through oxygen consumption did not differ appreciably from that lost through expiration of carbon dioxide. The cutaneous loss E_c was determined by fitting a sealed collar on the animal; the lower part of the body was then placed in a glass dessicator containing Drierite while the respiratory tract emitted water vapor directly to the external atmosphere; weight measurements were made at intervals of 4 to 12 hr. In some of these periods, there was no urinary loss and the respiratory term E_r could be obtained as $(N - E_c)$. Finally, the quantity U was estimated as the remainder term in Eq. (8.15).

TABLE XXV

WATER BUDGETS OF THREE AUSTRALIAN LIZARDS (ML WATER LOSS PER 100 G OF BODY WEIGHT PER DAY)[a]

Air temper- ature (°C)	Species	No. of speci- mens	Total loss N	Urine loss U	Skin evapo- ration E_c	Respiratory evaporation E_r
20	A. ornatus	7	2.98	0.53	1.72	0.73
	G. variegata	17	4.82	0.39	3.08	1.35
	S. labillardieri	7	7.97	1.99	2.44	3.54
30	A. ornatus	6	3.52	0.61	1.71	1.20
	G. variegata	7	6.59	0.25	3.64	2.70
	S. labillardieri	5	15.97	0.26	4.29	11.42

[a] Dawson et al. (244).

Table XXV summarizes the water budget for three species of lizards at 20 and 30°C. The principal differences are as follows:

(a) Water loss is greater through the skin than through the mouth for species A and G but not for S.

(b) Total water loss N is greater at 30°C than at 20°C for all species, and this is mainly due to an increase in E_r.

(c) Species S has the largest water loss of the three types of lizards.

Although this study was undertaken in a controlled laboratory environment, the results have ecological implications. The natural habitats of the three species are different, with the water balance of each lizard being the one most appropriate for its environment.

Getz (245) has estimated the water budget of the redback vole. The measurements were made in a laboratory also but in this case the specimens were fed. In addition, distilled water was supplied in a vial with an L-shaped drinking tube, from which values of water consump-

tion were obtained by weighing every 10 days; a correction for evaporative loss from the tube was estimated by weighing a similar device,
located nearby out of reach of the animals. Urine was collected in a pan
of mineral oil under the cages. The experiments were performed at 25°C
and at relative humidities of 95, 50, and 5%. Measurements of the two
largest terms in the water budget are given in Table XXVI, which shows
that relative humidity had little effect.

TABLE XXVI

WATER CONSUMPTION AND URINE PRODUCTION OF THE REDBACK VOLE (G/G-DAY)[a]

Relative humidity	No. of samples	Water consumption	Urine production
95	2	0.50	0.49
50	5	0.40	0.35
5	2	0.44	0.37

[a] Getz (245).

Total evaporative water loss $(E_c + E_r)$ of the voles was obtained over
6–8-hr periods by placing each animal in a dessicated chamber, through
which air was circulated at the rate of 250 cm^3/min. Average losses
were 0.0123, 0.0121, and 0.0146 g water per square centimeter of surface
area per 6 hr at 25, 28, and 33°C, respectively. There is no way of determining whether the water budget equation balanced because the measurements were not simultaneous; the activity of the animals was less in
the dessicated chamber than in the environment associated with the
values in Table XXVI.

Most physiological studies of water budgets are undertaken in the
laboratory. The primary interests are twofold; first, to examine the
effect of extreme stress caused by moisture deficiences, and second, to
determine the adaptive mechanisms that have permitted some organisms
to live comfortably in a desert environment. An ostrich is able to survive
because of its ability to drink salt water (246); the Australian desert
mouse, on the other hand, conserves water in its kidneys, being able to
gain weight on a diet of dry seed without drinking water (247).

An important consideration in both water and heat losses from an
organism is the ratio of surface area to volume. For a cube of length L,
the area is $6L^2$ and the volume is L^3, giving a ratio of $6/L$. For a sphere
of radius R, the area is $4\pi R^2$ and the volume is $4/3\pi R^3$, yielding a ratio
of $3/R$. These examples suggest a general rule that the evaporation and
heat losses per unit volume increase with decreasing size. Other things
being equal, a small insect loses water and cools more quickly than a

large animal. The size of an organism may be either an advantage or a disadvantage; in a very hot climate, a protective mechanism has probably evolved, such as the almost impermeable covering that prevents desiccation of an insect.

For a given volume, the sphere has the smallest surface area of any object. Thus, the igloo is the most efficient structure for heat and moisture conservation in a cold climate.

8.5. METABOLISM

Metabolism is a general term for the chemical activity taking place in living organisms, permitting cells to survive, organize, and grow. The metabolism of a complex structure such as man results from the metabolic activities of all the individual organs and tissues. When there is an optimum equilibrium condition among the parts, the body is said to be in a state of *homeostasis*.

The metabolic rate depends on the activity of the organism and on the state of its environment, the two factors often being interrelated. The metabolic rate of a resting organism is known as its *basal metabolism*. (More energy is consumed when a man is exercising vigorously than when he is sleeping.) A warm-blooded animal is called a *homeotherm;* a large fraction (as much as 80%) of the metabolic energy is used merely to regulate the body temperature. A reptile, on the other hand, is a *poikilotherm*, and more of the energy is available for motions of the creature, although some is still required to heat the body. Vegetation, too, is a relatively inefficient consumer of available energy (sunlight), less than 10% being used for creation of new plant material.

Some cells are able to survive or grow in the absence of oxygen (*anaerobic* mechanisms); these include some bacteria and fungi. Most living things, however, generally require oxygen to release energy (*aerobic processes*), the waste product being carbon dioxide. The CO_2 is diffused to the environment through the stomata of leaves, from lungs of animals, or through the skin (usually less than 1% of that from the lungs, in mammals).

Energy conversion in plants is by photosynthesis and requires sunlight. The general equations are the following:

Photosynthesis (intermittent):

(8.16) CO_2 (from air) + water (from soil) + solar energy →
                                  carbohydrates + oxygen (released to the air)

Respiration (continuous):

(8.17)   carbohydrates + oxygen (from air) →
                    CO_2 (to air) + water vapor (to air) + combustion energy

The CO_2 flux required for photosynthesis is typically a much larger quantity than that released by respiration, although at temperatures above about 35°C, the reverse is true. A vegetative cover is therefore a source of CO_2 at night but a sink during the day. Lemon (248), for example, suggests CO_2 exchange rates for corn of 0.345×10^{-6} g/cm^2-sec and 0.045×10^{-6} g/cm^2-sec for photosynthesis and respiration, respectively, when solar radiation is about 1 ly/min. The difference between these two quantities has been called *net photosynthesis*, as distinct from *gross photosynthesis*. For each gram of assimilated CO_2, about 90 g of water vapor are transpired (249).

This metabolic process can be monitored in a number of ways, but none is completely satisfactory. A critical review of these techniques has been given by Larcher (134).

(a) The dry-weight increase in the crop can be obtained at periodic intervals (the *harvest* method). This is a "destructive" test and furthermore yields only average values over relatively long periods of time (of little value for verifying micrometeorological estimates).

(b) Measurements of vertical gradients of CO_2 can be used to infer the flux. Above the canopy, however, there is no way of distinguishing between soil and plant contributions to the flux. The estimation of vertical flux divergences within the stand is therefore necessary.

(c) A leaf or branch is sometimes sealed in a plastic bag, the gas being monitored with an infrared CO_2 gas analyzer. This has been used, for example, by Hodges (250) on conifers. In addition to the effect of the plastic covering on the microclimate, the method yields only net CO_2 values. Sometimes respiration is measured in the dark, and this is added to the daytime net flux. However, there is no reason to believe that daytime and nighttime respiration rates outdoors are the same.

(d) Leaf water potential is measured with an electric psychrometer, as described by Spanner (251).

(e) Part of a leaf is cut, dried, and weighed. After 2 hr, a second cutting is made. The difference in dry weight per unit area of leaf yields gross photosynthesis. The physiological effects of cutting are reduced by killing the phloem at the basal parts of the lamina with hot steam just before the start of the experiment. This is known as the *half-leaf* method (252).

(f) Relative stomatal aperture is estimated either directly by microscopic examination or indirectly with a porometer; Slatyer and Jarvis (253) have described a gaseous-diffusion porometer in which N_2O is diffused through a leaf *in situ*.

Plant physiologists have sometimes complained that the meteorologist is more interested in the upward flux of water vapor than in the downward flux of CO_2. Metabolism is more important than transpiration in productivity studies, although there are interactions which require the investigation of both H_2O and CO_2 fluxes. Historically, water vapor has been easier and less expensive to measure than carbon dioxide; however, most agrometeorologists now recognize the importance of the latter gas.

Dry-weight production is not necessarily a good predictor of the market value of a harvest; of greater importance are the quality, quantity, flavor, color, and size of the edible portions of the crop. Seasonal studies (Chapter 11), as well as investigations of total ecosystems, must therefore be undertaken also. Nevertheless, a microscale understanding of what takes place at the surface of an individual leaf is an important link in the chain.

Stewart (254) has emphasized that world food productivity is dependent on nitrogen fixation as well as on photosynthesis. They are both metabolic processes, but the former has received little attention. Biometeorologists should study the nitrogen balance and fluxes in the biosphere, although the experimental determination of nitrogen gradients is not without difficulty.

Turning now to metabolic processes in animals and human beings:

(a) The source of energy is carbohydrates rather than sunlight.

(b) The rate of heat generation is significant whereas in vegetation the combustion energy is negligible.

(c) The rate of heat generation is variable, increasing greatly during strenuous exercise.

(d) Some of the metabolic energy is consumed as work (muscular activity).

(e) Some control of the metabolic processes is possible (clothing, heated homes, air conditioning, good nutritional practices, etc.).

An average man at rest inhales about 250 ml of oxygen per minute and exhales about the same amount of CO_2. The rates quicken as physical activity increases. The oxygen is carried through the blood stream by hemoglobin and is used to convert carbohydrates into heat and work. The ratio of the volume of CO_2 given off to the volume of O_2 consumed is called the *respiratory quotient* and is about 0.85 (0.7 for a pure fat diet and 1.0 for a pure carbohydrate diet).

The basal metabolism of a human creates heat at a rate of about 2000 kcal/day (255); this daily output can double when heavy manual labor is being performed. In comparison, the basal rate for a mouse in summer is about 7 kcal/day, to which can be added about 1.7 kcal/day for activities such as feeding and escape from predators (256). The diurnal

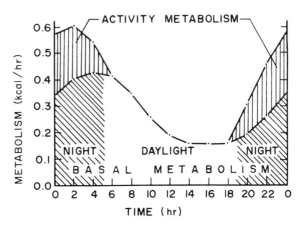

FIG. 62. Rate of metabolism of a 19-g wild mouse at various times on a June day (256).

cycle of metabolism of a field mouse is given in Fig. 62. The rate is highest at night because the animal is more active then and because the air temperature is lower; the basal rate generally increases in cool or very warm weather.

The metabolic rate is usually measured in cubic centimeters of oxygen consumption per gram body weight per hour or in cubic centimeters of CO_2 exhaled per gram body weight per hour. This is preferable to monitoring the diet because the energy equivalents of various foods are not sufficiently well known and because there is an uncertain lag between time of food consumption and time of energy conversion.

The physiology of a healthy organism in a favorable environment is reasonably well understood. The effects of very low or of very high temperatures (see Chapter 9) or of a lack of oxygen in the bloodstream can therefore be examined. This latter case is called *hypoxia* and is considered here. The condition occurs at high elevations or in the presence of some interfering gases. Carbon monoxide, for example, is absorbed by hemoglobin in the blood stream in preference to oxygen (in the ratio of about 300:1). Proper ventilation of automobile tunnels at high elevations is therefore of considerable concern, particularly in the critical case of many idling motors when traffic is blocked by an accident.

Miranda *et al.* (257) have developed a model for predicting the percentage of hemoglobin unsaturated with oxygen, as a function of elevation, concentration of CO, and exposure time. At 10%, a few individuals may experience slight headaches and dizziness; when the values rises to 35%, however, most people have severe headaches, dizziness, and other symptoms of hypoxia. The percentage P is given as

(8.18) $$P = 2.76 \exp(h/2100) + 0.0107 A C^{0.9} t^{0.75}$$

where

h = elevation above sea level in meters,
A = activity on a scale of 3 for rest, 5 for light activity, 8 for light work, and 11 for heavy work,
C = concentration of carbon monoxide in ppm,
t = duration of exposure in hours.

The effect of elevation alone is given by the first term on the right side of the equation. Its particular form arises from the fact that (a) the relation between atmospheric pressure and the percentage of unsaturated hemoglobin has been found experimentally to be exponential, and (b) atmospheric pressure decreases exponentially with height. The effect of carbon monoxide alone is given empirically by the second term on the right side of the equation; the assumption is made that there is no interaction with the first term.

Miranda *et al.* recommend that warning signs be displayed at the entrances to high-elevation tunnels, emphasizing also the added hazard of CO intake from cigarette smoking. Construction and maintenance workers should, of course, be screened for cardiopulmonary abnormalities at these heights.

8.6. ENERGY BALANCE

The energy balance of a flat solid surface impermeable to radiation is given by:

(8.19) $$Q_N = Q_G + Q_H + Q_E$$

where

Q_N = net radiation,
Q_G = conduction of heat through the solid,
Q_H = convective (sensible) heat transfer,
Q_E = evaporative (latent) heat transfer.

All values are to be obtained at or extrapolated to the surface itself. This equation is widely used in micrometeorology to study the energy balance of a large uniform area of bare ground or short grass (10).

When there is a change of state, the quantity Q_E must include the energy of freezing or thawing:

(8.20) $$Q_E = LE + L_f M$$

where

L = the latent heat of evaporation,
E = the evaporation rate,
L_f = the latent heat of fusion,
M = the melting or freezing rate in mass per unit area per unit time.

The energy balances of large uniform corn fields, forests, lakes, and glaciers must be formulated in terms of volume because solar radiation does not terminate at the upper surface but is attenuated downward. The appropriate equation is then

(8.21) $Q_{N_2} = Q_{G_1} + Q_{H_2} + Q_{E_2} + Q_S$

where Q_S is the rate of heat storage or loss within the volume and suffixes 1 and 2 refer to the lower and upper surfaces, respectively. In the case of vegetation, Eq. (8.21) includes additional small terms, such as consumption of energy by photosynthesis and generation by respiration. However, the magnitudes are less than the uncertainties in the estimates of Q_H and Q_E.

Finally, there are many three-dimensional problems; these are often difficult to solve, although the governing energy balance equations may be specified readily.

(a) The horizontal flat plate placed on the ground. Examples include an isolated irrigated field and an evaporation pan. Equations (8.19) and (8.21) are applicable at any given surface position but there are horizontal gradients to complicate the analysis.

(b) The suspended flat plate, which is used to model the heat balance of a leaf.

(c) The cylinder, which resembles a pine needle.

(d) Rectangular objects such as houses and factories. In these cases, the energy balance equation includes man-made sources of heat, as well as sinks created by air conditioning, ventilation, open windows, and infiltration through cracks.

(e) Insects, birds, animals, and human beings of all sizes and shapes, which are nevertheless often assumed to be cylinders or spheres. These forms of life move about in sun and shade and between windy and sheltered locations. The heat generated by metabolic processes is significant. In the case of man and domestic animals, furthermore, the energy balance is modified by indoor heating, air conditioning, or clothing.

For a living organism, the heat storage term Q_S is usually neglected; it must remain very small or life soon ends. The energy balance is then written

(8.22) $$Q_M = Q_B + Q_P = Q_N + Q_H + Q_E$$

where

Q_M = the heat energy from metabolic processes,
Q_B = the energy released by basal metabolism,
Q_P = that part of the muscular production of energy not utilized in work (see Eq. (8.24)).

If the body is prone, a conduction term Q_G must be added. The quantity Q_N has been transposed to the right side of the equation, as compared with Eqs. (8.19) and (8.21), to indicate that the major source of heat is metabolic rather than radiative. The term Q_M is always a gain while Q_N, Q_H, Q_E are usually losses; exceptions occur during exposure to intense radiation and during very hot weather when Q_N and/or Q_H may be gains with a compensating increase in Q_E through sweating or panting. The basal metabolism rate Q_B is approximately constant at 50 kcal/m² hr, and this is called 1 *met* (258, p. 33).

Equation (8.22) may be used to determine Q_P. However, this does not permit direct calculation of the muscular work rate W because of the variability of the efficiency f of the conversion of chemical energy CCE (255):

(8.23) $$W = fCCE$$

and,

(8.24) $$Q_P = (1 - f)CCE$$

Dividing Eq. (8.24) by (8.23),

(8.25) $$Q_P/W = (1 - f)/f$$

The quantity f averages about 10–20% for light work and up to 50% for strenuous exercise of short duration.

Although an independent estimate of Q_P is often required, the quantity is not readily determined. Monitoring of oxygen consumption yields values of Q_B when the body is at rest, and of the sum $(Q_B + Q_P + W)$ during activity. In order to obtain Q_P by this method, therefore, there remains the problem of determining W (or f). Nomograms and tables are available for estimating f, given the diet and the nature of the exercise. For more precise estimates, a volunteer is asked to operate a treadmill or bicycle, permitting direct monitoring of the work rate W.

Heat balance and water budget equations do not indicate *per se* the response of a living organism. Nevertheless, an understanding of the magnitudes of the sources and sinks of energy is a prerequisite to the study of stress–response relationships.

8.7. ENERGY CHAINS

Ohm's law states that

$$(8.26) \qquad \text{current} = \frac{\text{charge}}{\text{time}} = \frac{\text{potential difference}}{\text{resistance}}$$

The current per unit cross-sectional area is the electrical equivalent of a flux. Equation (8.26) has been used to model a number of transport processes, although the analogy is not strictly correct dimensionally. When values of any quantity (temperature, concentration, wind, etc.) are different at two points, the difference divided by the appropriate flux (of heat, matter, momentum, etc.) is defined as a *resistance*. In the case of heat conduction through a solid, for example, the rate of cooling Q in finite-difference form is,

$$(8.27) \qquad Q = k\frac{\Delta T}{\Delta z} = \frac{\Delta T}{R} \quad \text{or} \quad \frac{Q}{\rho c} = K\frac{\Delta T}{\Delta z} = \frac{\Delta T}{r}$$

where k is thermal conductivity (cal/cm sec°C) and K is thermal diffusivity (cm^2/sec). Equation (8.27) defines two resistances, R (°C cm^2 sec/cal) and r (sec/cm):

$$(8.28) \qquad R = \Delta z/k, \qquad r = \Delta z/K$$

Thus,

$$(8.29) \qquad R = r/\rho c = \Delta T/Q$$

The inverse of r has dimensions of cm/sec and is called a *transfer velocity* or *conductance*. Equation (8.28) indicates that a resistance depends on the thickness of the layer as well as on the physical properties of the medium.

When heat is flowing through several materials (plaster, wood, dead air space, insulation, and brick, for example), each layer having a temperature difference $(T_i - T_{i-1})$, and under steady-state conditions,

$$T_1 - T_0 = QR_1$$
$$T_2 - T_1 = QR_2$$
$$\cdot$$
$$\cdot$$
$$\cdot$$
$$T_n - T_{n-1} = QR_n$$

Adding these equations,

$$(8.30) \qquad (T_n - T_0)/Q = R = R_1 + R_2 + \cdots + R_n$$

Heat transfer from a house in winter has resistance in *series* as represented by Eq. (8.30) (through inner walls to outer walls) and also in *parallel* (heat losses through open windows and cracks, for example). If resistances are in series and there is an impenetrable barrier to heat transfer at one stage, the resistance at that point is infinite, as is also the sum of resistances. The flux is then zero.

Orlenko (259) has surveyed the methods used in the U.S.S.R. to estimate heat losses from buildings. The quantity Q is separated into two components: conduction, Q_c, and infiltration, Q_i. Infiltration is dependent on wind speed while Q_c is given by Eq. (8.30). Thus,

$$(8.31) \qquad Q = Q_c + Q_i = (T_o - T)/R_c + (T_o - T) f(u)/R_i$$

where T_o, T are the indoor and outdoor temperatures, respectively, and $f(u)$ is a function of the wind speed. Defining now a temperature T' such that

$$(8.32) \qquad Q = (T_o - T')/R_c$$

then the solution of Eqs. (8.31) and (8.32) is

$$(8.33) \qquad T' = T - (T_o - T) f(u) R_c/R_i$$

where T' is always less than T.

The quantity $f(u) R_c/R_i$ has been determined empirically for different types of structures and wind speeds. Orlenko presents a map showing isopleths of 0.1% probability values of T' over the U.S.S.R. The diagram has been used by the construction industry to divide that country into climatic regions (the effect of radiative heat exchange has been neglected). Orlenko calls T' an *effective temperature* but this term is reserved for another quantity to be discussed in Section 9.4.

The resistance R is often called *thermal insulation*, an expression found not only in engineering studies of building design but also in physiological investigations of the heat balances of men. In the latter case, R is frequently given in *clo* units, where 1 clo is the thermal insulation that maintains comfort in a resting man indoors at an air temperature of 21°C and relative humidity of less than 50%. If R_a, R_{cl} are the resistance of air and clothing, respectively, if the skin temperature is 33°C, if $Q_B = 50$ kcal/m² hr, and if 75% of Q_B is lost by convection, i.e., $Q_H = 38$ kcal/m² hr:

$$(8.34) \qquad R_a + R_{cl} = (33 - 21)/38 = 0.32 \quad °C \text{ m}^2 \text{ hr/kcal}$$

The value of R_a under these conditions has been found experimentally to be 0.14°C m² hr/kcal. Thus, the clo is defined as a resistance of 0.18°C m² hr/kcal. A related unit is the *Burton*, the vapor resistance of

1 cm of still air, the transfer of water vapor being by molecular motions only.

A resistance depends *inter alia* on the wind speed and the physical dimensions of the object; there is an increase in heat losses from a building when the wind speed increases and the outdoor temperature remains unchanged. Waggoner (260) has noted the similarity in the empirical expressions that have been given for air resistance of a leaf, of an insect, and of a man. After converting to common units of r (sec/cm):

(8.35) Leaf: $r_a = 2.3L^{0.3}u^{-0.5}$

(8.36) Insect: $r_a = 3.4L^{0.6}u^{-0.4}$

(8.37) Man: $r_a = 1.3L^{0.5}u^{-0.5}$

where L is a characteristic length and u is the undisturbed mean wind speed.

These three equations are used for water vapor as well as for heat transfer, i.e., the diffusion processes are assumed to be identical. To extend the resistance principle even farther, the entire soil–plant–air ecosystem can be modeled as follows:

$$(8.38) \quad \text{Flux} = \frac{P_1 - P_0}{r_r} = \frac{P_2 - P_1}{r_x} = \frac{P_3 - P_2'}{r_s} = \frac{P_4 - P_3}{r_v} = \frac{P_5 - P_4}{r_t}$$

where $(P_i - P_{i-1})$ is a potential difference and r_r, r_x, r_s, r_v, r_t are the root, xylem, stomatal, viscous, and atmospheric turbulent resistances, respectively (see Fig. 63) (261). There has been a recent controversy (262, 263) about the validity of relating resistances in the liquid and vapor phases. This seems to be partly a problem in dimensions. The quantities P_2 and P_2' are both potentials at the liquid/vapor interface but they are in different units. Thus, Eq. (8.38) cannot be rearranged and added to yield a simple relation between the flux and the potential difference $(P_5 - P_0)$ as in the case of Eq. (8.30). If r_L and r_V are liquid and vapor resistances in the same units, Cowan and Milthorpe (263) point out that Ohm's law is still a useful analogy, provided that the investigator realizes that equal changes in r_L and r_V do not yield equal changes in flux. For a specific case of $r_L/r_V = 2500$ at a temperature of 30°C, a change in r_V alters the transpiration rate 13 times more than an equal relative change in r_L.

The resistance concept has been valuable in the study of complicated physical systems. The "circuit diagram" for the heat balance of a room with a window and one outer wall is given as an illustrative example in Fig. 64; the details and notation can be obtained from the original paper by Parmelee (264). Another application is in the comparison of fluxes

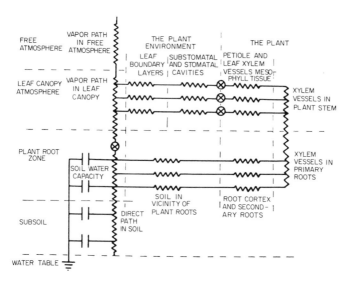

FIG. 63. Representation of pathways of water transport in the soil, plant, and atmosphere. Locations of phase change of liquid to vapor are distinguished by a cross (261).

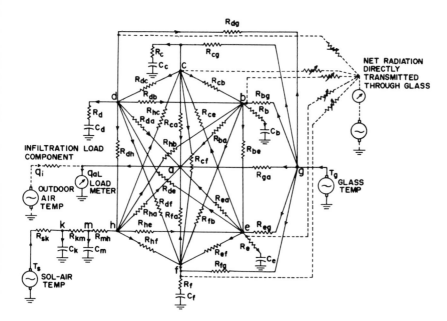

FIG. 64. Representation of the heat transfer from a room with three walls, floor, and ceiling adjoining similar rooms. The remaining wall and one window are exposed to the weather; for notation, see Parmelee (264).

of two different quantities through the same system. In the case of CO_2 flux into vegetation, for example, there is an additional mesophyll resistance not present in the case of water vapor flux. Waggoner and Zelitch (265) note that for tobacco, maize, and cotton at least, the partial closing of stomata reduces transpiration more than it reduces photosynthesis. In applications such as these, the resistances are expressed preferably in units of seconds per centimeter.

A casual survey of the literature reveals a wide variety of units. Engineers prefer to use a *heat transfer coefficient* h_H, which is the ratio of heat flux Q_H (langleys per unit time) to temperature difference $(T_2 - T_1)$ (°C). Thus,

(8.39) $h_H = Q_H/(T_2 - T_1) = \rho c/r = 1/R$

Similarly, a mass transfer coefficient is given by h_M, which is the ratio of flux to difference in mass. For water vapor,

(8.40) $h_M = E/(q_2 - q_1) = \rho/r$

When there is a flow of air $\bar{u}$ over a surface, with a flux of some quantity X perpendicular to the surface, a nondimensional *transfer* coefficient C_x is often used.

(8.41) $C_x = \text{flux}/(X - X_0)\bar{u} = 1/r_x\bar{u}$

where X_0 is the surface value of X and $\bar{u}$ is measured at some given distance away from the interface. Note that C_x and r_x are both dependent on the distance from the surface. In the case of vapor, heat, and momentum, respectively,

(8.42) $C_E = \dfrac{E/\rho}{(q_s - q)\bar{u}}$

(8.43) $C_H = \dfrac{Q_H/\rho c_p}{(T_s - T)\bar{u}}$

(8.44) $C_D = \dfrac{\tau/\rho}{(\bar{u} - 0)\bar{u}} = \dfrac{\tau}{\rho\bar{u}^2}$

The quantity C_D is called the drag coefficient and τ is the shearing stress.

If ν, α, and D are the molecular diffusivities of momentum, heat, and mass respectively, then

(8.45) $\text{Pr} = \nu/\alpha$ (the Prandtl number)
(8.46) $\text{Sc} = \nu/D$ (the Schmidt number)

When the surface itself is one point in the measurement of gradients, and the flow is turbulent, it is important to know whether the flux is from an aerodynamically smooth or a fully rough surface. In the former case, the transfer near the interface is determined by molecular viscosity; in the latter case, there is an additional *form drag*, caused by separation of the flow around bluff obstacles (Karman vortices in the ideal case).

For the smooth-surface turbulent case, it has been found that (266, p. 416)

$$(8.47) \qquad C_H \operatorname{Pr}^p = C_E \operatorname{Sc}^p = C_D$$

where p is an exponent having a value of about 2/3. This relation seems to apply also for a fully rough flow, provided that C_D is separated into viscous and form drag, C_v and C_f, respectively. Then,

$$(8.48) \qquad C_H \operatorname{Pr}^p = C_E \operatorname{Sc}^p = C_v$$

After calculating C_v, C_D from Eqs. (8.48) and (8.44), respectively, C_f may be inferred. Thom (267) has measured C_H, C_E, and C_D for an artificial leaf in a wind tunnel; he used three liquids, water, bromobenzene, and methyl salicylate, with differing molecular diffusivities. His data are in agreement with Eq. (8.48), as are some outdoor measurements by Barry and Munn (268) with iodine-131 and tritiated water vapor.

This brief introduction to the problem of fluxes from surfaces is described in more detail in engineering texts and handbooks (266). The objective is to achieve universal relationships and to investigate fluxes in complex systems.

Bryant (269) has provided an interesting study of the determination of working limits for a continuous atmospheric release of iodine-131 based on a maximum permissible concentration in milk of 0.4 mμc/liter. The resulting equation is

$$(8.49) \qquad Q = C/\bar{d}VTA$$

where

Q = the release rate of ^{131}I,
C = the maximum permissible concentration in milk,
$\bar{d}$ = the weighted mean dilution factor (determined from Pasquill's diffusion equation and his atmospheric stability categories) (see Section 13.2),
V = the transfer velocity to the grass (the inverse of resistance),
T = the effective mean life of ^{131}I on the grass,
A = the activity in milk resulting from unit deposit on grassland.

The methods used to estimate these various quantities are given in the original paper. This example has been introduced merely to illustrate the fact that attempts are being made to use physical methods to follow a substance through the various links of an ecosystem.

Another approach to the study of energy chains is by computer modeling. Although some of the initial or boundary conditions may not be known, they may be simulated to provide a prediction of conditions at some later time. This in itself is not a very productive exercise. However, the power of the procedure becomes apparent when the computer program is rerun, changing by a relatively large amount one of the variables or boundary conditions. The importance of each link in the ecological chain can then be evaluated separately. This is sometimes called a *sensitivity* analysis. Although subsequent verification is always necessary, the investigator is at least provided with some clues concerning interactions in a complex multisystem. As suggested by Sargent (270), it may be possible to use this method to predict the ecological implications of proposed major man-made weather modifications. Recent examples of computer simulations of complex systems have been given for the transpiration process by Woo *et al.* (271), for carbon monoxide concentrations in cities by Ott *et al.* (272), and for ecological population changes by Garfinkel (273).

Computer modeling frequently takes advantage of the *Monte Carlo method*, i.e., the selection of random numbers from a set having an appropriate standard deviation. This procedure may be introduced at one or more steps in the chain for two reasons:

(a) To investigate the stability of the solution to experimental noise, in which case random errors may be introduced deliberately.

(b) To simulate a link (such as population response to an environmental stimulus) that can only be expressed in probabilistic terms.

An example of the Monte Carlo method is given in Section 10.5.

Finally, energy chains may sometimes be modeled with an electrical *analog*. When the governing differential equations are identical to those describing the flow of electricity through a complex non-steady-state circuit, the biometeorological process may be simulated on a small analog computer. Herrington (274), for example, was able to predict the temperature field in a tree stem, using a differential equation that describes not only heat transfer but also the potential in a noninductive electrical cable: the analysis revealed that there was a heat flow resistance between bark and xylem.

8.8. DIFFUSION AND VENTILATION

Air flows obey conservation equations. The volume of air per unit time entering one end of a pipe must equal the volume leaving at the other end. Similarly, the pollution emitted from a chimney per unit time must be accounted for by wind transport, turbulent diffusion, absorption at the ground, and chemical decomposition in the atmosphere.

A critical survey of diffusion theory has been given by Pasquill (16); the topic is so diverse that the methodology cannot be summarized in a few pages without misleading the reader. Only a single example of diffusion, illustrative of a "balance" equation, is given here.

Suppose that pollution is emitted from a chimney at a constant rate Q. Suppose also that an upper inversion based at height h (higher than effective chimney height) is preventing the pollution from diffusing upward above level h. If the mean wind speed in the lower layer is $\bar{u}$, with x, y the downwind and cross-wind distances, with the chimney as origin,

$$(8.50) \qquad Q = \bar{u}h \int_{-\infty}^{\infty} C \, dy$$

where C is the ground-level concentration of pollution at any given point (x, y). The model assumes that there is no loss of pollution at the ground and that x is sufficiently large for complete vertical mixing of the pollution up to height h.

Assuming next that the cross-wind distribution of pollution is Gaussian with standard deviation σ,

$$(8.51) \qquad C/C_0 = \exp(-y^2/2\sigma^2)$$

where C_0 is the center-line ground-level concentration. Substituting Eq. (8.51) into (8.50) and integrating,

$$(8.52) \qquad Q = \bar{u}hC_0\sigma(2\pi)^{1/2}$$

A balance equation has yielded a result of engineering value.

The ventilation of buildings is also a very diverse topic. A basic parameter is *air changes per unit time*, although this does not imply that there is a complete renewal of fresh air in one air change.

If C_A is the outdoor concentration of a pollutant, C_I is the indoor concentration at some initial time, and C is the indoor concentration at any later time, with $C_I > C > C_A$ and C_A constant, then,

$$(8.53) \qquad dC/dt = S(C - C_A)$$

where S is defined as the *air-change rate* in units of t^{-1}. Thus,

$$(8.54) \qquad (C - C_A)/(C_I - C_A) = \exp(-St)$$

The quantity $\tau = 1/S$ is called the *ventilation time constant* of the building. The quantity is similar in concept to that defined by Eq. (2.14).

For a 32-m^3 unheated experimental structure, Georgii (275) has found values of about 0.5 hr^{-1} for the ventilation time constant.

Because there are so many types of buildings and materials, ventilation systems are not easy to design. As an aid, for example, air leakage values through windows (volume of air per unit time per unit length of sash perimeter) are determined experimentally and published in building guides. Lundqvist (276) emphasizes the growing importance of such studies, because buildings are becoming more airtight. Paradoxically, the older Danish schools (constructed before 1904) have a much greater air-change rate than modern ones, and contain fewer "dead zones" where air-change rates are low: the "stale" air in these zones may diminish the power of concentration of students.

9/PHYSICAL METHODS: ILLUSTRATIVE EXAMPLES

9.1. INTRODUCTION

The physical methods described in Chapter 8 frequently yield only partial solutions to problems. For example, experimental data on the heat balance or metabolism of a human do not indicate whether the subject is comfortable. In general, as the heat balance departs further from "normal," the chances of stress increase, but the "comfort zone" is not easy to specify or to explain. There is no obvious reason, for example, why the preferred body temperature should be exactly 37°C (98.6°F) (but see Section 9.4).

Three types of measurements are usually made when undertaking an experiment in human biometeorology. First, the physical environment is monitored, permitting calculation of the stresses. Second, the subjects are asked to describe their reactions. Finally physiological measurements (of blood pressure, pulse rate, etc.) are taken. These measurements should be made preferably at the time that stress is applied, but the subject may not be aware of any unusual influences, and it may be many weeks before an examination is made. In the short-term laboratory situation on the other hand, the response may be conditioned by the presence of the sensors. Despite these and other difficulties, the threefold approach is recommended wherever possible.

The bibliography on stress–response interactions is very great and could not be summarized easily. This chapter includes only four illustrative topics—environmental stress arising from humidity, cold, heat, and ecological competition. The book by Landsberg (258) is recommended for further reading.

9.2. THE EFFECT OF HUMIDITY ON HUMAN COMFORT

The popular expressions, "damp cold," "damp heat," and, "it's not the heat but the humidity," indicate a common belief that the moisture content of the air has an effect on human comfort. This may be examined physiologically by questioning a group of volunteers or by making standard measurements of skin temperature and so forth. With very low relative humidities, however, the lips become dry and the skin may itch; this can affect the subjects' judgments of a comfortable condition. Furthermore, if the face is exposed and the arms are bare, there is uncertainty about which part of the body is inducing a feeling of discomfort.

An alternative method suggested by Wong (277), will be described as an example of a purely physical approach to a biometeorological problem. Wong considered an outdoor condition of cloudy skies, or an indoor environment with no strong radiant heat sources. The evaporative heat loss Q_E from the body is given by

$$(9.1) \qquad Q_E = (q_s - q_a)/R_v$$

where

q_s, q_a = specific humidities at the skin and in the air,
R_v = the resistance to vapor transport (see Section 8.7).

Equation (9.1) assumes that R_v consists of a sum of resistances in series, i.e., that clothing covers the entire body.

The convective heat transfer Q_H is given by

$$(9.2) \qquad Q_H = (T_s - T_a)/R_h$$

where

T_s, T_a = skin and air temperatures,
R_h = the resistance to convective heat transport.

Dividing Eq. (9.1) by Eq. (9.2),

$$\frac{Q_E}{Q_H} = \frac{(q_s - q_a)R_v}{(T_s - T_a)R_h}$$

For a wet-bulb thermometer in a moving air stream, R_v/R_h is a constant, C, independent of temperature or relative humidity. For a fully clothed man with wet skin, on the other hand, $R_v/R_h < C$. The ratio R_v/CR_h has been called the *moisture permeability index* P (nondimensional) by Woodcock (278). It depends on the type of clothing but is supposed to be independent of temperature and humidity. Thus,

(9.3) $$Q_E/Q_H = PC(q_s - q_a)/(T_s - T_a)$$

where $0 < P < 1$. In the case of a human homeotherm, neglecting extreme ambient temperatures which induce physiological responses such as sweating or shivering, the effect of a change in relative humidity is assumed to be modeled by Eq. (9.3).

For a wet-bulb thermometer ($P = 1$) kept at a temperature of 35°C, the magnitude of the ratio Q_E/Q_H is given in Table XXVII for a series

TABLE XXVII

RATIOS OF EVAPORATION TO SENSIBLE HEAT TRANSFER FOR A WET BULB
THERMOMETER MAINTAINED AT A CONSTANT TEMPERATURE OF 35°C (95°F)[a]

Ambient air temperature		Q_E/Q_H		$\dfrac{(Q_E/Q_H)_{dry} - (Q_E/Q_H)_{sat}}{(Q_E/Q_H)_{dry}}$
°C	°F	Dry air	Saturated air	
26.7	80	10.1	3.8	0.62
21.1	70	6.1	3.4	0.44
15.6	60	4.3	3.0	0.31
10.0	50	3.4	2.6	0.22
4.4	40	2.8	2.4	0.15
−1.1	30	2.3	2.1	0.10
−6.7	20	2.0	1.9	0.06
−12.2	10	1.8	1.7	0.04
−17.8	0	1.6	1.6	0.02
−23.3	−10	1.4	1.4	0.01
−28.9	−20	1.3	1.3	0.01
−34.4	−30	1.2	1.2	0
−40.0	−40	1.1	1.1	0

[a] Wong (277).

of ambient air temperatures and for 0 and 100% relative humidity. Also included (in the last column) is the percentage change in this ratio when the humidity changes from 0 to 100%. The following points are of interest:

(a) Provided that air temperature is lower than surface body temperature, as is the case for all values given in Table XXVII, there is evaporation even at 100% relative humidity ($Q_E/Q_H \neq 0$ in column 4).

(b) For temperatures below freezing, a change of relative humidity from 0 to 100% has hardly any effect on the ratio Q_E/Q_H.

(c) As the air becomes warmer, the effect of a humidity change becomes increasingly important. For a temperature of 26.7°C (80°F), the ratio Q_E/Q_H changes from 10.1 to 3.8 when the humidity increases from 0 to 100%. Assuming constant Q_H, therefore, the heat loss by evaporation decreases as the humidity rises.

These results apply to a wet-bulb thermometer and their relevance to a human being must be considered. First, clothing reduces the magnitudes of the ratios in columns 3 and 4 by the fraction P when the human skin is wet; however, the relative changes (in the last column of Table XXVII) are unaffected. Second, only part of the skin is likely to be wet; the ratios are therefore reduced even further but again the percentage changes are not affected. Finally, the case is considered of constant room temperature with a change from 0 to 100% relative humidity. Unless thermal resistance is decreased by removing clothing and/or increasing ventilation, the evaporative heat loss is reduced by about 50% (see Table XXVII), creating a sensation of warmth. Thus, damp heat seems hotter than dry heat.

Such drastic humidity changes never occur. For a more modest increase from 50 to 100% at a temperature of 26.7°C (80°F), Wong has computed that there would be a decrease of about 34% in the heat loss from the human body.

As the ambient air temperature decreases, the effect of a humidity change becomes less and less, becoming negligible below 0°C. However, care must be taken not to extrapolate these results to very cold (or very warm) weather. If water vapor condenses in the clothing, two additional processes must be considered. In the first place, heat is released by the change in state from vapor to liquid and/or to ice. Second, the thermal conductivity of the clothing is increased, which in turn increases the convective heat losses from the body. Burton and Edholm (**279**, p. **65**) emphasize that heat transfer through a system with changes of state is not an easy problem. This complexity is important not only in the design of arctic clothing but also in the insulation of buildings.

9.3. WIND CHILL

Another limiting case of considerable importance is wind chill. When temperatures are well below zero and strong winds are blowing, convective heat losses from the body become painfully great.

Siple and Passel (**280**) determined the rate of freezing of water in a plastic cylinder (15 cm in length and 6 cm in diameter) on 89 occasions in Antarctica, expressing their results in kilocalories per square meter of exposed surface per hour. They fitted their data to an empirical equation:

(9.4) $Q_H = (10 \sqrt{u} + 10.45 - u)(33 - T_a)$

where u is the mean wind speed in meters per second and T_a is the air temperature in degrees Celsius; thus, $(33 - T_a)$ is the difference between air and normal skin temperature. The quantity Q_H is called a *wind-chill index*. Equation (9.4) may be used to normalize various combinations of u and T_a to a reference wind speed of 2.2 m/sec (5 mph), i.e., T_a is converted to a temperature that would yield the same heat loss Q_H if the wind were only 2.2 m/sec. A nomogram may therefore be prepared giving *wind-chill temperature* (of a plastic cylinder of water!) as a function of u and T_a. This information is often given to the radio and press during cold waves. For example, when u is 22.3 m/sec (50 mph) and T_a is $-18°C$, the wind-chill temperature is about $-45°C$. Values of Q_H can be as large as 1000 (very cold) to 2500 kcal/m² hr (intolerably cold).

Court (20) has reexamined Siple and Passel's data. He believes that a better empirical fit is given by the equation

(9.5) $Q_H = (10.9 \sqrt{u} + 9.0 - u)(33 - T_a)$

Court also suggests a physical approach, using the theory of heat transfer from cylinders. The nondimensional numbers of importance are the Reynolds (ud/v) and the Nusselt $(h_H d/k)$ numbers, where d is the diameter of the cylinder, v, k are the viscosity and thermal conductivity of air, and h_H is the transfer coefficient of Eq. (8.39). This leads to the prediction,

(9.6) $h_H = cu^m/d^n$

where c is a constant and m and n are exponents determined experimentally. Gates (6, p. 107) quotes values of $\frac{1}{3}$ and $\frac{2}{3}$, while Buettner (281) suggests $m = 0.5$ and $n = 0.5$. Other objections to Eq. (9.4) as a wind-chill index have been raised by Burton and Edholm (279, p. 110). First, the effect of wind on heat loss cannot be considered without reference to the amount and type of clothing being worn. Second, human comfort is more likely to be determined by sensations in exposed extremities such as the face and ears rather than by the heat balance of the entire body. Even when the body is protected by clothing, the heat transfer rate from a finger (small cylinder) is greater than that from a leg (large cylinder). Finally, blowing snow that melts on the skin increases the evaporational cooling rate; according to Massey (282), this sometimes results in frostbite.

Despite these objections, the concept of wind chill seems to be of qualitative value. Wilson (283) has related records of frostbite in Antarctica to values of Q_H obtained from Eq. (9.4). The results (Fig.

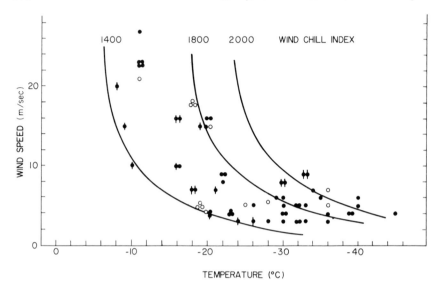

TEMPERATURE (°C)

Fig. 65. Recorded cases of frostbite in Antarctica in relation to temperature and wind speed: (○) changing meteorological conditions (the mean of the extremes in temperature and wind has been used); (◆) observations of frostbite during sledding (wind speed estimated); (●) stationary meteorological conditions. The curved lines represent wind chill indices of 1400, 1800, and 2000, respectively (283).

65) suggest that exposed skin begins to freeze at a wind-chill index of 1400 kcal/m² hr.

The resistance R_a of heat transfer from clothing to air is given by

(9.7) $Q_H = (T_{cl} - T_a)/R_a$

Burton and Edholm (279, p. 50) have quoted the following experimental result:

(9.8) $1/R_a = 0.61(T_a/298)^3 + 0.19u^{1/2}(298/T_a)$

where T_a is in degrees Kelvin (298°K is room temperature), u is in centimeters per second, and R_a is in clo units. Equation (9.8) combines the losses by radiation and convection and is almost independent of T_a. Although there may be a vapor pressure gradient, the reasonable assumption is made that no evaporation is taking place at the surface of clothing. Values of R_a obtained from Eq. (9.8) for various wind speeds are given in Table XXVIII. This illustrates the fact that the insulation of clothing becomes less effective in stronger winds.

The cooling power of air is of interest at temperatures above freezing also. Measurements are made in some countries with the *frigorimeter,* a blackened sphere of about 7.5 cm diameter, the surface of which is

TABLE XXVIII

VALUES OF R_a FOR VARIOUS WIND SPEEDS[a]

R_a (clo units)	0.8	0.6	0.5	0.4	0.3	0.2	0.15	0.1
u (m/sec)	0.2	0.3	0.6	1.1	2.2	5.3	10.1	22.8

[a] Burton and Edholm (279).

maintained at a high temperature (about 36.5°C). This is sometimes called a *globe* thermometer. Flach and Mörikofer (284), for example, have prepared a climatology of outdoor cooling power (in millicalories per centimeter squared per second) for a number of locations, mostly in Switzerland. The measurements integrate the effect of convection and radiation but exclude evaporational cooling (except during periods with precipitation or dew). Thus, a perfect correlation cannot be expected with physiological data.

Globe thermometers are used indoors also. In this environment, there has been a trend towards smaller spheres in order to reduce the time constant. Lidwell and Wyon (285) have designed a blackened bead 2 mm in diameter enclosed in a thin spherical polyethylene shell 0.1 mm thick and 25 mm in diameter; the time constant is about 1 min.

Measurements from the outdoor globe thermometer are sometimes included in an index of human comfort. A typical expression is

$$(9.9) \qquad T_{WGT} = 0.7T_W + 0.2T_G + 0.1T_D$$

where T_{WGT} is called the wet bulb globe temperature index, and T_W, T_G, and T_D are the wet bulb, globe, and dry bulb temperatures, respectively (286). Equation (9.9) has no physical basis.

Pugh (287) has described a laboratory study of the circumstances surrounding the death by exposure of a youth on the "Four Inns Walk" in Derbyshire, England in March 1964. The original conditions were simulated by a healthy volunteer, 21 years of age, who underwent tests at a temperature of 9.4°C, a relative humidity of 50% and a maximum wind speed of 4 m/sec in a climatic chamber. He was dressed in a jacket with hood, woolen sweater, wool-cotton shirt, underwear, jeans, one pair of socks, shoes, and gloves. Measurements were made while the subject was resting and after exercising on a bicycle ergometer at a work rate of 300 kg m/min. A comparison was also made between conditions of dry and wet clothing.

Because this particular study illustrates many biometeorological principles, it will be described in some detail. Mean skin temperature T_s was calculated from 2–4 readings made with a thermocouple at each of 13 points. Mean body temperature T_b was assumed to be the sum of $\frac{2}{3}$

TABLE XXIX

LABORATORY MEASUREMENTS OF THE HEAT BALANCE OF A MAN[a,b]

No.	Conditions	T_a (°C)	T_s (°C)	T_R (°C)	Q_M (kcal/hr)	W (kcal/hr)	Q_E (kcal/hr)	Q_S (kcal/hr)	Q_H (kcal/m hr)	R_a (clo)	R_{cl} (clo)	Air movement (m/sec)
2	Rest dry	9.4	30.58	37.50	127	0	34	+2	48	1.0	1.49	0.06
3	Rest dry	9.4	26.93	37.48	115	0	30	+51	69	0.4	1.00	1.1
4	Work dry	9.4	31.24	38.20	300	41	46	+22	118	—	—	0.13
5	Work dry	9.4	26.44	38.28	370	41	54	−8	134	0.23	0.48	4.0
6	Work wet	9.4	21.10	38.06	450	43	71	—	168	0.23	0.17	4.0
7	Work wet	9.4	21.50	37.76	458	54	77	0	171	0.23	0.17	4.0

[a] Pugh (287).

[b] T_s, mean skin temperature; T_R, rectal temperature; Q_M, metabolism; W, mechanical work; Q_E, evaporative heat loss from lungs and skin plus heat loss warming inspired air; Q_S, rate of stored heat, a plus sign denotes cooling; Q_H, nonevaporative heat loss through clothing; R_a, air insulation; R_{cl}, clothing insulation. The values are the means of the last two observations during each period.

rectal and $\frac{1}{3}$ skin temperature. Stored heat loss Q_S was obtained from the equation,

$$(9.10) \qquad Q_S = \Delta T_b \cdot W \cdot c \cdot 60/t$$

where

ΔT_b = the change in the mean body temperature in time t min,
W = the body weight in kilograms,
c = the specific heat of tissues (0.83 kcal/kg).

The sum of the resistances of clothing R_{cl} and air R_a was obtained from the formula,

$$(9.11) \qquad R_{cl} + R_a = (T_s - T_a)/0.18Q_H$$

The factor 0.18 converts cgs to clo units.

The value of R_a was obtained from Eq. (9.7), while Q_H was estimated as the remainder term in the heat balance equation (8.22). This required a knowledge of the magnitude of Q_M. The CO_2 concentration of expired air was monitored, in addition to the oxygen intake. During bicycle ergometry, the work rate of 300 kg m/min was divided by 427 to obtain its thermal equivalent.

The results are given in Table XXIX for several experiments. Although Pugh emphasizes that some of the estimates are uncertain, the trends are clear. Insulation is less effective in strong winds than light ones but is least effective with wet clothing. The resistance R_{cl} dropped to 0.17 clo in the last case, only 15% of the value for dry clothing in light winds with the volunteer at rest. At the same time, average skin temperature decreased to about 21°C (down to about 8°C in the foot). Pugh concludes that the cause of the "Four Inns Walk" tragedy was exposure to wind in wet clothing, although there may have been a contributing psychological factor. The knowledge that shelter and dry clothes were not immediately available would have an adverse effect on the walkers. Runners in training, on the other hand, endure similar conditions deliberately, knowing that they can step indoors at a moment's notice.

9.4. HEAT STRESS

The first defense of a human being against a warm environment is dilation of the blood vessels, resulting in an increase in the flow of blood near the skin. This warms the surface but maintains the deep body temperature. In a cold environment, on the other hand, the blood flow decreases. The ranges over which these mechanisms are effective are called the zones of *vasomotor regulation* against heat and cold.

The second line of defense against heat stress is through sweating, panting, and evaporational cooling. If temperatures of the air and of surrounding walls or other solid objects are higher than skin temperature, the body gains heat by convection and radiation; evaporation is the only remaining cooling mechanism.

Although it is easy to state these simple principles, it is difficult to develop heat-stress indices, particularly when they are based solely on meteorological variables. There are a number of physiological reasons for this. The air temperature at which sweating begins varies greatly, even among a group of people at rest in the same environment. The activity and clothing of a man are important, as well as the factor of climatic adaptation. The sweat rates of newcomers to the tropics are almost invariably greater than those of the local inhabitants, although part of the explanation for this may be psychological. Portig (288) suggests that the immigrant must adapt to a new culture as well as to a new climate; the weather is sometimes blamed for other difficulties in readjustment. Nevertheless, laboratory studies (289, p. 115) have shown that when volunteers performed light work under very hot conditions for 4 hr a day, their endurance showed daily improvement for at least 2 weeks. Part of the change is physiological, the sweat containing less salt after two weeks. On a much shorter time scale, a cold drink of water on a hot day reduces the sweat rate temporarily and therefore does not cool at all; hot tea is preferable.

Marinov (290) has suggested that the fact that body temperature is constant at $37°C$ can be explained by recalling that man originated in equatorial regions. Assuming an air temperature of $30°C$ and a vapor pressure of 33 mb, he considers the heat balance of a man at noon for three different body temperatures T_B. When $T_B = 30°C$, the body is not in equilibrium with its environment and must gain heat. When $T_B = 44°C$, the body must lose heat. Finally, when $T_B = 37°C$, equilibrium can be achieved over a range of metabolic rates through regulation of the sweat rate. If man had originated in another environment, the sweating mechanism might have evolved differently [see also Burton and Edholm (279, pp. 11–14)].

Many heat stress indices have been proposed. Some are based on the subjective reactions of volunteers to various combinations of temperature, humidity, wind, and radiation. However, human perception of thermal comfort is not fully understood. Chatonnet and Cabanac (291), for example, suggest that shivering may be the *cause* of the impression of cold during a fever; the patient may "feel" cold although in fact his body temperature is high.

Other indices employ physiological measurements. Tromp (292), for example, has suggested several chemical tests. As noted by Lee (293), however, the joint effects of separate stresses are not necessarily additive. For example, pulse rate rises as air temperature increases or as the level of physical exercise goes up; however, if air temperature and work rate are increased simultaneously, the change in pulse rate "bears no predictable relation" to the sum of the separate increases. Hence, pulse rate is not a specific reproducible index of heat stress.

One index that has been used for many years by heating and ventilating engineers is the *effective temperature*, "the temperature at which motionless, saturated air would induce, in a sedentary worker wearing ordinary indoor clothing, the same sensation of comfort as that induced by the actual conditions of temperature, humidity, and air movement" (294, p. 192). This index does not include the effect of radiant heat and must be adjusted empirically for moderate to strong winds. Values of this index were obtained by analyzing the group response of volunteers as they entered one controlled chamber from another. The air flow rate was 13 cm/sec (25 ft/min). First impressions were used in order to achieve reproducibility; after a few minutes of exposure, the subjects were not able to distinguish moderate gradations of comfort. The results are given in the form of a nomogram, Fig. 66 (289, p. 117) which is in degrees Fahrenheit. For long exposures, Fig. 66 overestimates the effect of humidity on comfort (289, p. 123). Another uncertainty arises because the nomogram represents only average response for a population, and there are regional differences. Figure 67 (289, p. 124) shows the percentage of people in several cities that felt comfortable at various effective temperatures. The peak frequencies occur at higher effective temperatures in San Antonio than in Minneapolis. Other differences exist because of age, state of health, and so forth.

The upper part of the nomogram in Fig. 66 can be fitted empirically by a straight line. When the units are in degrees Fahrenheit, the equation takes the simple form (295),

$$(9.12) \qquad ET = 0.4(T_D + T_W) + 15$$

where T_D and T_W are dry- and wet-bulb temperatures, respectively. In this case, the effective temperature is called the *humidity index* and is a measure of discomfort indoors or in the shade outdoors with light winds.

Hendrick (296) has added the effects of wind and solar radiation to produce a more complete comfort index. As Hendrick himself notes, however, "in reality, there can be no precise numerical measure of as subjective a concept as comfort."

Quite a different approach, the use of physiological indicators, has been recommended, although again a one-to-one relationship with

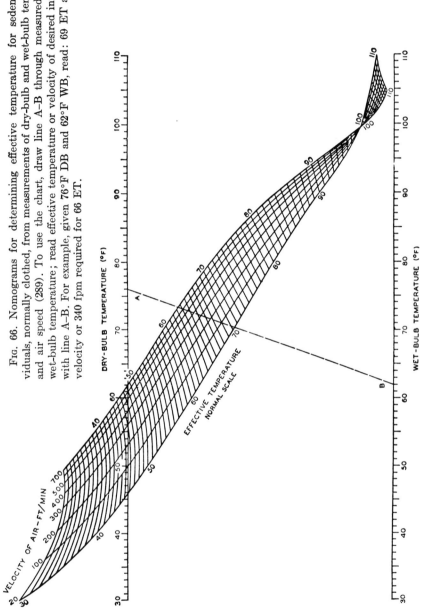

Fig. 66. Nomograms for determining effective temperature for sedentary individuals, normally clothed, from measurements of dry-bulb and wet-bulb temperatures and air speed (289). To use the chart, draw line A–B through measured dry- and wet-bulb temperature; read effective temperature or velocity of desired intersections with line A–B. For example, given 76°F DB and 62°F WB, read: 69 ET at 100 fpm velocity or 340 fpm required for 66 ET.

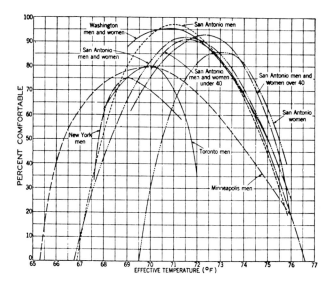

FIG. 67. Relation between effective temperature and the percentage of observations indicating comfort in several cities (289).

comfort is not to be expected necessarily. One such indicator of some merit is sweat rate. An integrated value for the entire body may be obtained from weight change measurements at hourly intervals. The sweat rates over relatively small surface areas (about 12 cm²) can also be monitored, by attaching a ventilated capsule to the skin to measure the change of humidity. In this kind of experiment, the assumption is made that the sweat rate is not affected by the presence of the capsule. Van Beaumont and Bullard (297) used a small electric heating coil with thermostatic control to keep the skin temperature in the capsule at some constant elevated value. In one experiment when the air temperature was 26.5°C, two capsules on the left forearm were maintained at 34.5 and 39.1°C. Initially there was no sweating in the heated skin areas or elsewhere. Moderate work was then begun on a bicycle ergometer. Sweating started first under the warmer capsule and was subsequently about three times as great as that under the cooler one. This is too large a difference to be explained by physical factors, i.e., by a greater vapor pressure gradient, and indicates a physiological thermoregulatory mechanism.

Lee and Vaughan (133) used the portable chamber shown in Fig. 68 to study the effect of solar radiation on the heat balance of man. The walls and roof were covered with Mylar film, and the vertical fluxes of downward and reflected upward short-wave radiation were measured within the chamber. In some experiments, sunlight was shielded with

FIG. 68. Portable chamber designed for studies of the effect of solar radiation on heat stress (133).

metal shades. Humidity, temperature, and ventilation controls were provided.

Acclimatized volunteers walked on a treadmill at 94 m/min for 2 hr. The solar radiation loading was estimated by multiplying the vertical flux of solar radiation by the area of a volunteer's shadow on a horizontal surface. While the subject was standing still with arms down, his shadow area was traced on heavy paper, cut out, weighed, and divided by the weight of 1 m² of paper. This value was multiplied by 1.05 to allow for the extra area of swinging arms when on the treadmill. The reflected solar radiation loading was obtained by assuming that the body was a cylinder, of length equal to the height of the man, and of radius such that the cylinder cast a shadow equal in area to that cast by the man. Finally, the albedo of the clothed human body was estimated to be about 0.3. The short-wave radiation loading Q (kcal/m² hr) was then given as,

$$(9.13) \qquad Q = 0.7(Q_T + Q_R)/A$$

where Q_T and Q_R are the downward and upward fluxes (in kilocalories per hour), respectively, on the subject and A is his surface area (in square meters). This loading was of course not distributed equally but was greater on the sunny side. Replications of Q were obtained during low and high ventilation rates (45 and 152 m/min), for several air temperatures, and for volunteers wearing desert tan uniforms on some occasions and in shorts only at other times; the vapor pressure was kept constant. Control experiments were carried out with solar radiation shielded.

These data were related to mean sweat rate (in grams per hour), obtained from the loss in nude body weight during the 2-hr period, and to the mean sweat evaporation (in grams per hour) obtained from the gain in weight of clothes. For equal sweat rates, air temperatures were compared in the sunny and shielded cases, and regression equations determined. The results were then expressed in terms of ITER (the increment of air temperature equivalent to the radiation load), as given in Table XXX. The ITER is less with shorts than with uniforms and is less with fans than without. Furthermore, the estimates of ITER from measurements of body sweat loss are not greatly different from those of clothing sweat rate.

The concept of ITER is a logical extension of an approach recommended by Lee and Henschel (298, 299), based on an earlier proposal by Belding and Hatch (300) for a heat strain index:

(9.14)

$$RS = \frac{\text{evaporative cooling required to balance a heat stress}}{\text{maximum possible evaporative cooling in the same environment}}$$

A value of zero indicates no stain, and other values are obtained

TABLE XXX

INCREMENT OF AIR TEMPERATURE EQUIVALENT TO RADIATION LOAD ITER[a]

Index	Conditions	Air temperature in sun (°C)	Equivalent air temperature when sunlight shielded (°C)	ITER (°C)
Mean sweat evaporation	Shorts, fans	44.6	37.8	6.8
	Shorts, no fans	45.1	37.4	7.7
	Uniformed, fans	45.1	37.3	7.8
	Uniformed, no fans	47.2	37.4	9.8
Mean sweat loss	Shorts, fans	44.1	37.8	6.3
	Shorts, no fans	44.6	37.4	7.2
	Uniformed, fans	45.8	37.3	8.5
	Uniformed, no fans	45.9	37.4	8.5

[a] Lee and Vaughan (133).

empirically. Lee and Henschel call this the *relative strain* RS. They also express the index in terms of resistances and gradients using the notation of Section 8.7. Assuming a skin temperature of 35°C and a saturated vapor pressure of 44 mm Hg, Eq. (9.14) retains its generality when written in the form,

$$(9.15) \qquad RS = \frac{Q_M(R_{cw} + R_a) + 5.55(T_a - 35) + Q_N R_a}{7.5(44 - e_a)}$$

where R_{cw} is the resistance of wet clothing in clo units, e_a is the vapor pressure of air, and the other terms are as defined in Section 8.7. Because Eq. (9.15) contains too many variables to be interpreted quantitatively, Lee and Henschel have considered first a simplified case in which:

(a) A man is indoors with wall temperatures equal to air temperature (no radiant heat sources).
(b) The man is dressed in a light business suit and is walking at 26.8 m/min ($R_{cw} = 0.4$).
(c) The air flow is 30.5 m/min ($R_a = 0.4$).

Equation (9.15) then becomes,

$$(9.16) \qquad RS = \frac{10.7 + 0.74(T_a - 35)}{(44 - e_a)}$$

This relation may be displayed in the form of a nomogram.

Second, adjusted values of RS were obtained for various values of Q_M, Q_N, and wind speed, after careful examination of published physiological data on heat stress: the American Society of Heating, Refrigerating, and Air-Conditioning Engineers (ASHRAE) sponsors a continuing program of research in this field. For strong solar radiation loadings, the ITER data in Table XXX provide the additional adjustments. Lee and Henschel prefer to apply Eq. (9.16) (by making appropriate corrections to T_a and e_a for nonstandard conditions) rather than to prepare a series of nomograms based on special solutions of Eq. (9.15).

Finally, values of RS must be related to degrees of discomfort. This requires laboratory data, using questionnaires and "comfort votes" on samples of healthy, aged, sick, and acclimatized populations. Lee and Henschel have presented some tentative relations based on published studies but they recognize that more experimental data are required.

As a footnote, *equivalent temperature* T_E should be mentioned. It is the temperature that air would have if all its water vapor were condensed, the latent heat being used to heat the air, i.e.,

$$(9.17) \qquad T_E = T_a + Lw/c_p$$

where L is latent heat of condensation and w is the mixing ratio. The quantity T_E is used widely in the design of air-conditioning facilities, and engineers often seek climatological information on its frequency distribution.

9.5. ECOSYSTEM COMPETITION

Populations are hardly ever in equilibrium with their environments. Even in an undisturbed habitat far from civilization, there are short-term weather cycles as well as long-term climatic trends, and each periodicity has an associated time constant for population response; in addition, a catastrophe such as a forest fire or a flood may produce irreversible effects. More important is man's intervention, including pollution of the air, soil, and water, and the deliberate manipulation of plant, animal, bird, and insect life.

For a rather abrupt environmental change, there is likely to be a readjustment in population densities and species, called *succession*. Assuming a new constant environment, the populations approach asymptotically a steady-state plateau, termed a *climax*. The concept of climax has been the subject of controversy for a long time but it is now clear that the idea has qualitative but rarely quantitative significance.

Ecologists search for isolated populations which can be studied as a closed ecosystem, on a remote island, for example. In this connection, care must be taken that the scientists and their equipment do not themselves disturb the environment.

A study of an isolated community has been described by Maycock and Matthews (301). They have found a willow thicket in a river valley in northern Ungava, Quebec, about 500 km north of the tree line. Although the preliminary field work raises more questions than it answers, the authors recognize the importance of the site as a long-term outdoor laboratory. Because the willows exist near the threshold limits for survival, the area can be used to seek interrelationships among growth-limiting factors. For example, an opportunity is provided for study of arctic species that normally grow on the flat open tundra but are shaded in the willow thicket.

Ecologists are interested in the growth or decline of populations, and in the productivity per unit area. The two topics are closely related, both depending on environmental conditions and on the competition for available food and water. Productivity, for example, is affected by the density of population; closely spaced seedlings may not yield as large a harvest as a thinned crop, although there must clearly be some optimum spacing beyond which the yield per unit area declines.

Many empirical equations in ecology are rather similar to the one which models the growth of leaf rust (302, p. 56):

$$(9.18) \qquad dx/dt = kx(1 - x)$$

where

    x = the rusted fraction of the leaf area,
    t = time,
    k = a constant, provided that the susceptibility of the host does not change.

Thus, the growth of the disease to epidemic proportions is related to the rusted fraction that is producing fungi as well as to the nonrusted fraction that is not yet infected. Waggoner (303) notes that k is not constant but is a function of temperature, dropping to zero in both very hot and very cold weather. Experimental data on potato blight fungi suggest a parabolic curve for k, with a maximum value at about 20°C. Thus,

$$(9.19) \qquad k = k_{\max} - c(T - 20)^2$$

where c is a constant. In a very simple model, a sinusoidal form is assumed for mean daily maximum temperature during the summer,

$$(9.20) \qquad T = a_0 + a_1 \sin t$$

Equations (9.18)–(9.20) lead to climatological predictions concerning epidemics that can be tested for different values of the constants.

Summer temperatures are not smooth functions as modeled by Eq. (9.20); furthermore, other weather variables such as humidity are important. Waggoner therefore develops another model that employs consecutive 3-hourly observations of temperature, relative humidity, sunshine, and leaf wetness (the leaf is assumed to be wet if rain has fallen within the 3-hourly period during daylight hours or within 6 hr at night); a computer is programmed to make a separate decision on the growth or decay of infection after receiving each 3-hourly observation. The logic of the decision process is rather crude but Waggoner concludes rightly that meteorological conditions are rarely suitable for the development of a potato blight epidemic.

An equation similar to Eq. (9.18) is used to predict the yield of a healthy crop (304):

$$(9.21) \qquad \frac{d \ln w}{dt} = \lambda(t)[1 - w/W]$$

where

w = the dry weight at any time t,
W = the dry weight at harvest time,
λ = the coefficient of growth and is a function of time.

The equation may be solved to yield,

(9.22) $$w = W/(1 + ke^{-T})$$

where

$T = \int\lambda(t)\,dt$,
k = a constant of integration.

For multicomponent systems, the relations are naturally more complex. A controversy exists as to whether a population is self-regulating, i.e., whether it is density dependent only, or whether it responds to environmental stresses. For insects at least, it seems that both mechanisms may play a part. In the Engadin Valley of Switzerland, for example, a "strong" form of the gray larch budmoth is dominant at times when populations are low, because of a high reproductive capacity and an ability to disperse (305). With increasing densities, a "weak" form is favored as a result of resistance to granulosis virus. At even higher densities, however, the weak form is attacked differentially by parasites, thus reversing the cycle. For other insects, density-dependent oscillations are small and weather cycles may control the growth and decline of populations, although there is often uncertainty about the nature of the response. In a study of insect populations in various parts of Canada, Watt (306) concludes that the annual fluctuations are more variable in mild than in harsh climatic regions. He suggests that through the process of natural selection or of adaptation, insects in severe climates are less sensitive to weather than those in more temperate regions.

The other major ecological question is productivity, which is being investigated intensively during the IBP (International Biological Program). Relevant terms are *biomass* (in units of grams per square meter) and *productivity* (dimensions of grams per square meter per day). The ratio of the two is in units of days and is called the *turnover time* (307).

In a study of grassland ecology in California's San Francisco peninsula, McNaughton (308) employs two additional indices. The *dominance index* is the percentage of the total standing crop contributed by the two most common species, and the *diversity index* is the total number of species: estimates of both were obtained by random sampling of the area. McNaughton finds, as might be expected, that dominance varies inversely as diversity. In addition, however, productivity varies linearly as dominance but inversely as turnover time; the greater the number of species, the less productive is the grassland.

There are many indices of diversity, a fact that reflects a fundamental difficulty in statistical representation of a region containing many species, each population tending to occur in clumps rather than in a random fashion. Three of these indices are given in Eqs. (9.23)–(9.25).

(9.23) $$d = (S - 1)/\ln n$$

where S is the number of species and n is the total number of individuals in the sample (309).

(9.24) $$D = \sum_{i=1}^{S} (n_i/n) \ln(n_i/n)$$

where n_i is the number of individuals in the ith species (310).

(9.25) $$\Delta = \left(\sum_{i=1}^{S} n_i^2 \right)^{1/2}$$

where Δ is a third diversity index (311).

The relative merits of the three equations will not be debated. It seems clear that they are all incomplete and that the way to proceed is by spectral and cross-spectral representations. The methods described in Chapter 7 may be applied, replacing time by distance, i.e., using wavelength instead of frequency as the abscissa in Fig. 54, p. 141.

10/SYNOPTIC APPLICATIONS

10.1. THE SYNOPTIC METHOD

The synoptic approach has frequently been rewarding in biometeorology. Whereas monthly and seasonal averages of the meteorological elements smooth away important extremes, the synoptic method helps to clarify nonlinear interactions. Total rainfall during a growing season, for example, may be sufficient in quantity but the timing may be unsatisfactory. In some agricultural regions, precipitation is important only during two or three short but critical periods associated with sprouting, spraying, and harvesting.

At coastal, valley, and urban locations, prestratification of data according to the synoptic pattern is recommended, and a useful index in many cases is the geostrophic or gradient wind. The development of a lake breeze, for instance, depends *inter alia* on whether the "regional" wind is blowing off land, off water, or parallel to the coastline. Sometimes, as in the case of long-distance travel of pollen and spores, the pressure patterns at heights of a few kilometers must also be examined. There is a further class of problems in which the persistence of certain synoptic anomalies is critical; for instance, an anticyclone may be associated with high levels of pollution or an extreme risk of forest fires.

Schroeder (312) examined the weather charts for days in May over a 4-year period when foresters in Wisconsin and Michigan had reported critical burning conditions. Composite surface pressure maps were drawn for North America for Days —2, 0, +2, where Day 0 was the first day of critical burning conditions. The charts showed a dominant high-pressure center that moved from just west of Hudson Bay southeastward to Virginia in 5 days.

Once this synoptic pattern has been identified, the forest meteorologist can begin meaningful case studies. He should, in particular, determine whether the Hudson Bay high pressure area is a necessary and sufficient condition for a serious fire hazard.

A long-term goal of biometeorologists has been the development of a meaningful classification of air masses and synoptic patterns. Arakawa and Tawara (313), for example, have tabulated the frequency of air-mass types in Japan, while investigators in Europe have studied *Gross-wetterlagen* (large-scale weather features), particularly in a medical context. Landsberg (314, pp. 222–232) has provided a useful summary of these studies.

This approach has not been as successful as had been expected. Air masses are not in fact horizontally homogeneous, and Creswick (315) has stated that the concept of a widespread body of air with uniform properties is not realistic, even in the free atmosphere. Near the ground, variations in the meteorological elements are often greater across an air mass than across a front.

Air-mass and synoptic-pattern classifications should be as simple as possible. Lamb (316), for example, has used only six classes in his studies of weather patterns in the British Isles: anticyclonic, cyclonic, northwesterly, northerly, easterly, and southerly. In some studies, days that do not need well-defined criteria should be discarded.

For microclimatic investigations, Wilmers (317) recommends the use of only four synoptic classes—radiation type, squall weather type, cyclonic type, and neutral weather type—based on three factors—net radiation, advection, and vertical motion (subsidence or convection in the middle troposphere). After the climatological frequencies of these classes have been established for a region, it is possible to determine whether a relatively short set of micrometeorological observations is representative of the long-term microclimate.

10.2. STAGNATING ANTICYCLONES: POLLUTION POTENTIAL FORECASTING

High pollution potential is defined as a synoptic-scale meteorological condition which, given the existence of multiple sources, is conducive to the occurrence of high concentrations of pollution. The definition is in terms of meteorological factors only, and high pollution *potential* may occur when air quality is excellent in the Canadian arctic far from civilization, for example.

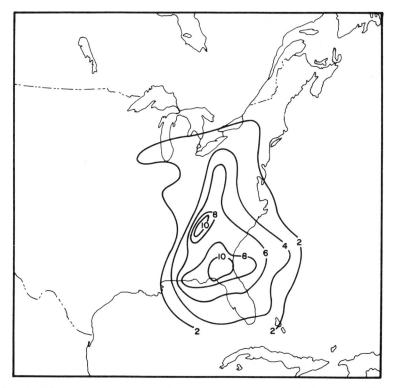

Fig. 69. Total number of stagnation cases (7 or more days) in the period 1936–1965 (318).

Synoptic research in the United States over a period of more than 10 years has led to the development of objective methods of pollution potential forecasting. The National Meteorological Center in Suitland, Maryland prepares daily 36-hr alerts, a subprogram of the regular computer weather prognostics. The area of interest is the continental United States, with a limiting macroscale resolution of about 1.9×10^5 km² (75,000 miles²); the lines delineating advisory regions are bands approximately 160 km (100 miles) wide.

Poor air quality traditionally has been associated with spells of weak horizontal and vertical ventilation (light winds and temperature inversions). These conditions are most likely to occur in stagnating anticyclones, the climatology of which has been studied by Korshover (318) for the eastern United States. Figure 69 shows isopleths of the total 30-year numbers of stagnation cases lasting 7 days or more, using the criteria that the geostrophic wind must be less than 7.5 m/sec (15 kt)

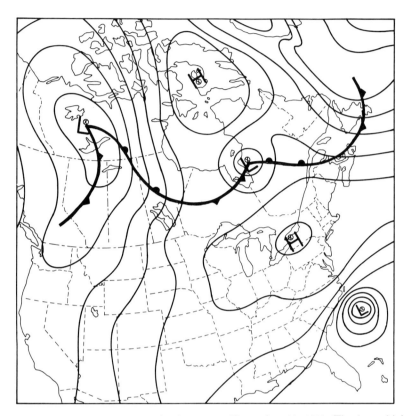

Fig. 70. Surface weather map for 7 PM EST, November 30, 1962. The large high-pressure area over Lake Ontario contributed to poor air quality throughout the region.

and that frontal and precipitation areas be excluded. The maxima in the southeastern United States are associated with westward extensions of the Bermuda anticyclone and with high pressure areas originating in northern Canada that become stationary along the Atlantic seaboard. A typical weather pattern is shown in Fig. 70; on that occasion in Toronto, a national football championship game had to be postponed because of smog.

Seasonally, stagnation cases are most likely in autumn, with a secondary maximum in spring. A prolonged spell of light winds is not so common in summer but when such a situation does arise, pollution levels may not necessarily be high* because emissions are less at that time of year and because vertical ventilation is greater (longer hours of daylight and

* An exception is photochemical oxidant smog.

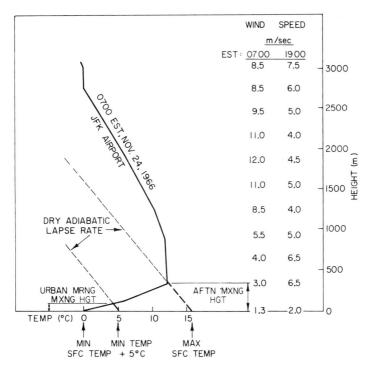

Fig. 71. Vertical temperature and wind speed data for John F. Kennedy Airport, New York, 0700 EST, November 24, 1966. The method of calculating afternoon and urban morning mixing heights is indicated (320).

stronger insolation). An important meteorological variable for pollution potential forecasting is therefore the *mixing height*, the layer of air near the ground through which pollution is stirred and diluted by convection. Using the standard network of rawinsonde stations, Holzworth (319, 320) has studied the climatology of afternoon mixing heights in the United States. An actual graphical calculation is shown in Fig. 71 (320).

During clear skies and light winds at night, there is often a radiation inversion (zero mixing height) in the country but an adiabatic lapse rate near the ground in cities (due to the heat island). The morning urban mixing heights can be estimated from rural rawinsonde ascents and urban surface temperatures, as indicated in Fig. 71. The afternoon mixing height is assumed to be about the same in the city as in the country.

The other variable of importance in pollution potential forecasting is the average wind speed through the mixing height. If this layer is very thin, a strong wind has the same effect on pollution transport as a lighter wind associated with a larger mixing height. The *ventilation coefficient* is defined as the product of mixing height and average wind speed. For small urban areas, wind direction is also important.

The present criteria and operational procedures at the United States National Meteorological Center include the following (321):

(a) The observed 1200 GMT rawinsonde data are used to calculate urban morning mixing heights and average wind speeds for these layers. To adjust for the urban heat island, 5°C is added arbitrarily to the surface minimum temperatures reported at 1200 GMT. Two of the criteria for an air pollution alert are a morning mixing height of less than 500 m and an average wind speed of less than 4 m/sec.

(b) The mixing heights for the first afternoon of the forecast period are obtained from the forecast maximum temperatures in the way indicated in Fig. 71, i.e., the dry adiabatic line through the maximum temperature is followed upward to its intersection with the 1200 GMT profile. The mean wind speed is estimated from forecast wind speeds at fixed levels for 00 GMT using an interpolation procedure.

(c) Estimates of mixing heights and winds for the second afternoon are based on an empirical method developed by Miller (322), which relates these quantities to forecast temperatures and winds at standard pressure levels.

(d) The product of forecast afternoon mixing height and wind speed (the ventilation coefficient) must be no greater than 6000 m²/sec and the predicted wind speed must not exceed 4 m/sec on both afternoons.

(e) There must be no significant precipitation or frontal passages during the 36-hr period.

(f) The affected area must be at least as large as a 4 deg square of latitude and longitude.

(g) Forecast 1500-m winds within the area must average less than 10 m/sec.

(h) The temperature change at 1500 m during the previous 12 hr must must be $\geq -5°C$.

These conditions are deliberately restrictive. When an alert is forecast, it is almost always verified meteorologically, although some poor air quality situations occur which are not predicted. In any event, the local synoptic meteorologist does not rely entirely on the alert but may modify the forecast on the basis of his knowledge of local conditions.

The approach has been developed to provide warning of spells of poor horizontal and vertical ventilation. Existing models of diffusion fail in this limiting case of light and variable winds when some of the most serious pollution conditions occur. The method has its shortcomings as might be expected of any objective system. The limiting meteorological conditions are chosen arbitrarily, such as the 500 m restriction on mixing height at 1200 GMT, although any other value would be equally difficult to justify. Also, there has been a natural tendency to use currently available "products" of the National Meteorological Center, rather than to begin from first principles.

An alternative approach is to forecast air quality. Each city is a special case, however, with its own mesoclimate and areal distribution of pollution sources. Numerical models are being developed (323) in which the atmospheric diffusion equations are solved on large computers for each chimney and area source. A more modest approach is to seek empirical regressions between air quality and meteorological measurements at each sampling location. These regressions must include the effect of persistence, which is difficult to model synoptically. Williams (324) has suggested that actual air quality for the previous 24-hr period be used as one predictor.

Velds (325) has related the daily SO_2 concentrations in Rotterdam to the Grosswetterlagen defined for Central Europe by Hess and Brezowsky (326). There are 28 circulation types in this classification, but only three of them are correlated significantly with high SO_2 values in the Netherlands.

HM: A high pressure belt over Central Europe.
HN_a: A high pressure area over the northwestern part of the North Sea, producing north to northeast winds in the Netherlands.
Sz: A southerly circulation.

Velds has also investigated the effect of persistence of these three circulation types on SO_2 concentrations. In the case of type HM, the frequency of days with mean SO_2 values greater than 300 $\mu g/m^3$ was 12% for runs of 2–6 days but rose to 52% for longer spells.

10.3. DISTANT TRANSPORT OF GASES AND PARTICLES

The story of wheat rust is not only an interesting chapter in history, but it also illustrates man's attempts to understand the role of the atmosphere in the transport of plant pathogens. Wheat stem rust normally spends part of its life cycle on wheat and part on the barberry bush. To begin the chain (see Fig. 72), *aeciospores* are released in prodigious numbers from infected barberry in spring. These spores do not attack other barberry but instead use wheat as host. If sufficient moisture

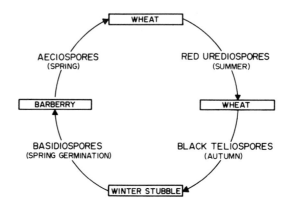

Fɪɢ. **72.** The life cycle of wheat stem rust. Barberry acts as alternate host.

is available, the spores germinate in a few hours. After an incubation period of a week or so, red *urediospores* are emitted into the atmosphere to reinfect the fields of wheat; this is a repetitive stage which may intensify the damage exponentially. In the autumn, however, a change occurs: there is a release of black *teliospores,* capable of withstanding winter cold. These germinate in spring to produce the fourth type of spore, *basidiospore,* which infects barberry rather than wheat to complete the cycle, as illustrated schematically in Fig. **72.**

As early as 1660, a law was passed in northern France requiring the destruction of barberry in wheat-growing areas (327, p. 21). In 1917 at a meeting of cereal pathologists at Madison, Wisconsin (327, p. 21), a resolution was approved requesting the President of the United States to establish a commission to consider "the desirability of eradication of all cereal rust-bearing strains of the barberry in the United States." By the mid-1920's, most of the barberry had, in fact, been eliminated in the American and Canadian prairies but nevertheless, outbreaks of wheat rust did not diminish. The possibility of continental-scale transport of spores was considered, but initial studies eliminated this hypothesis (327). The barberry aeciospores do not develop south of 39° N latitude (except in mountains) because they cannot survive in very hot weather. Wheat urediospores, on the other hand, are winter-killed north of Texas and also do not withstand summer heat in the southern United States. How then was it possible for the barberry to act as alternate host?

Subsequent investigations showed that the red urediospores (the summer repetitive spore of the cycle) were transported by contimental-scale air flows, southward to Texas in autumn and northward to the prairies in spring. Wheat rust could therefore propagate from year to year in the complete absence of barberry. The use of spore traps in the 1920's and 1930's provided detailed maps of the distant travel of spore clouds during 2–3-day periods of persistent southerly (or northerly) winds between the Gulf of Mexico and the Canadian prairies. Today it is possible to predict infestations from synoptic weather charts, although few warning services have been established, largely because of the lack of effective methods of chemical control; instead, the emphasis has been on the development of rust-resistant strains (327).

There are other instances of double-host relationships; apple rust, for example, requires the presence of both apple and red cedar trees. Because there is no repeating stage, however, the two species must grow within a few kilometers of each other. Apple scab, on the other hand, needs no alternate host; infected dead leaves release *ascospores* after rainfall in spring whenever temperature conditions are favorable. A critical factor in this case seems to be the duration of leaf wetness (328).

Countless numbers of pollens, spores, and molds are transported over distances of thousands of kilometers. During large-scale southerly flows over North America in spring, for example, tree pollen may cause hay fever before budding has commenced locally. As another instance, Hogg (329) has used the trajectory method (Section 5.7) to determine the source regions for spores reaching southern England. Spore catches are small when the air has come from the Atlantic but high when trajectories are from Spain.

In recent years, scientists at the Rothamsted Experimental Station in cooperation with the Meteorological Research Flight (153, 178) have been measuring both the vertical and downwind concentrations of spores. Hirst and Hurst (178) emphasize that distant transport is governed by synoptic weather, not by climate. Three-dimensional data obtained during particular synoptic situations are more likely, therefore, to assist in understanding the interplay of physical and physiological mechanisms than is a study of monthly or seasonal spore-trap concentrations. Hirst and Hurst find that spore populations are nearly constant with height in the daytime convective mixing layer over southern England but decrease almost logarithmically upward in an inversion. Penetrative convection, however, creates intermittent pockets of high concentrations within upper temperature inversions.

The British Isles consitute a volume source for pollens and spores. Hirst *et al.* (153) have described the results of aircraft sampling over the North Sea, on occasions when winds were southwesterly, for cladosporium and pollens released mainly during the day, and for damp-air spores emitted at night. The results obtained from one flight are given in Fig. 73, which has several noteworthy features:

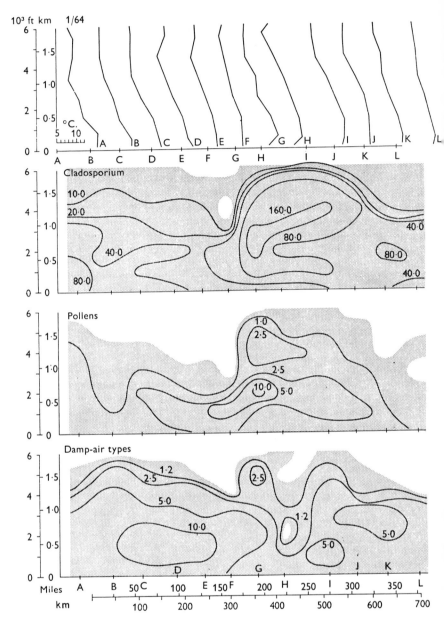

Fig. 73. Vertical temperature profiles (top) and isopore diagrams of *Cladosporium*, pollen, and damp-air types. The abscissa scale is down-wind distance over the North Sea from the English coastline (153).

(a) Concentrations are not highest at the surface but at heights of 500–1000 m, indicating a loss of particles to the sea.

(b) The height of maximum concentrations is 300–600 m lower for pollens than for cladosporium, a fact that may be explained by differences in gravitational settling rates.

(c) The maximum concentrations are not at the English coast but out to sea, at a downwind distance of about 350 km for the two daytime types, and about 150 km for the nocturnal damp-air spores. The data were obtained during the period 0945–1215 GMT, July 16, 1964, and a suggested explanation is that the clouds of cladosporium and pollens at downwind distances of 350 km originated over England on July 15, whereas the damp-air spores were released 12 hr later and therefore had not traveled as far. In the case of cladosporium, the diagram suggests that a new cloud is forming at the English coastline, and that the cloud emitted on July 14 can still be detected at a distance of 650 km and height of 500 m.

The patterns over the North Sea in Fig. 73 support the view that urban pollution does not remain within local or national boundaries. For

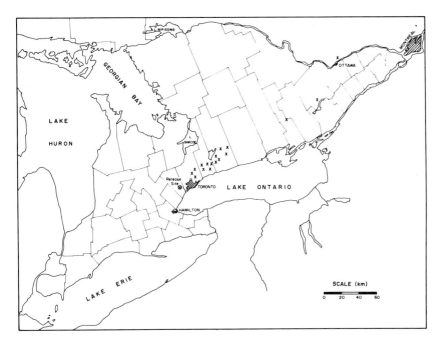

Fig. 74. Map showing recovery locations of balloons released on October 23, 1967 from a site near Toronto. Winds were southwesterly throughout the area (331).

the argon-41 point source at a height of 105 m at the Brookhaven National Laboratory on Long Island, New York, Petersen (330) has used an aircraft to sample the plume over the Atlantic on a day when winds were northwest and the lapse rate was neutral. The invisible cloud was still detectable at a downwind distance of 290 km, the outer limit of aircraft sampling. Pollution peaks in Greenland and elsewhere have been traced back to industrial midlatitude source regions, using the trajectory method.

With the growth of the "megalopolis," synoptic-scale transport of pollution is becoming increasingly important. That pollution can travel long distances is illustrated in Fig. 74 (331). As part of the 1967 "Cleaner Air Week" program in Peel County, Ontario, there were daily releases of helium-filled balloons designed to travel at a nominal height of 200 m above the ground. Tags were attached, many of which were returned subsequently by mail. On October 23, there were 17 returns, all directly downwind from the release site (winds were southwesterly on that day). One balloon was found in the Ottawa area, and presumably others traveled even farther but were not retrieved.

10.4. DISTANT TRAVEL OF INSECTS AND BIRDS

Because insects and birds can fly, their motions cannot be predicted from meteorological considerations alone. On one particular occasion in Ceylon, for example, an observer watched three species of migrating butterflies traveling in three different directions (332). Nevertheless, group behavior is often influenced by the wind and by turbulence.

The diamond-back moth appeared in great numbers in 1958 not only in Britain but also as far north as the Faeroe Islands and Iceland. Although the pest is indigenous to Europe, the fact that the population explosion in Britain occurred suddenly, beginning first along the east coast, is rather convincing evidence for an immigration. The annual totals caught in a trap at Kincraig in northern Scotland are given in Table XXXI (333). In 1958, the catch increased greatly during the night of June 29–30, as it did also in a trap on the Island of Bornholm in the Baltic. Shaw (333) has examined the antecedent trajectories. He suggests that the moths originated in eastern Europe and he speculates that the initial buildup occurred over a vast area. The ecological reasons for the exodus are not known but synoptic weather patterns were favorable for a westward movement towards Britain. Migrations of this particular moth on such a scale are relatively rare, the only other recorded infestations in Britain being in 1891 and 1914.

TABLE XXXI

NUMBER OF DIAMOND-BACK MOTHS CAUGHT AT KINCRAIG, SCOTLAND, 1955–1959[a]

Year	Number caught	
1955	61	Trap began on 23 May
1956	100	
1957	66	
1958	2201	
1959	29	Trap ceased 23 August

[a] Shaw (333).

As another illustration (334), the 1962 invasion by the small mottled willow moth in southern England has been studied by the method of trajectories. The moths appear to have begun their travel in Morocco four days earlier; although the analysis is somewhat uncertain over such long travel times, the method provides a useful preliminary hypothesis.

The aphid is an insect that has been observed to travel great distances. For example, myriads of spruce aphids arrived over a large area of the ice cap of Spitzbergen on August 7, 1924 (335); the nearest source region is the Kola peninsula, 1200 km to the southsoutheast.

Johnson (336) has examined the vertical distributions of aphids at Cardington, England, using traps mounted on a balloon cable at 6 levels up to heights of 600 m. The sampling time was 1 or 2 hr, and the temperature difference was measured between 0.9 and 600 m. On many occasions, the vertical profiles of aphids could be fitted to a power law:

$$(10.1) \qquad C/C_1 = (z/z_1)^{-p}$$

where C is the concentration of aphids at height z, and p is an exponent that varies from about 0.3 to 2.0. The smallest values of p occurred during superadiabatic conditions when turbulent mixing was greatest. There were usually two waves of aphids during the day, one in the morning and the other in the afternoon. The morning concentration peak was often lagged with height, by as much as $2\frac{1}{2}$ hr, suggesting inefficient vertical mixing; the afternoon peak occurred almost simultaneously throughout the profile. However, Johnson is not certain whether there was, in fact, a delay in upward transport or whether several populations were involved.

The desert locust is a larger insect whose motions are of considerable concern. A single swarm may contain 10^9 locusts and consume 10^8 kg of plant material per day (337, p. 4); the crop damage is therefore counted in terms of millions of dollars. The process of swarm formation is not fully understood, mainly because it occurs suddenly and perhaps

1000 km from the area where crop damage subsequently takes place. However, Roffey and Popov (338) were able to observe a buildup in the Tamesna area of the Niger and Mali Republics in 1967. During September and October particularly, large numbers migrated into the region. They flew during the 3-hr period after sunset, traveling a distance of about 9 km per evening but not displaying any group behavior. There were also shorter daytime flights (13–20 m) towards preferred patches of vegetation. Soil moisture was important in the egg-laying and multiplication stages; as soon as the ground had dried to a depth of about 10 cm, the females ceased to deposit eggs or they moved to moister sites such as automobile tracks. By the end of October, the population was about 16 times that of the original number of immigrant parents, due, in part, to the fact that natural predators were absent. As the population increased, the locusts underwent physiological and behavioral changes, such as the development of black markings and a tendency towards the formation of gregarious bands.

Kennedy (339) distinguishes between the seemingly purposeful low-level straight-line flights of individual locusts and the passive transport across continents of swarms carried by the stronger upper winds at heights of 500–2000 m. In Africa, for example, rainfall occurs in different regions at different times of the year, and the locusts "tend to be delivered to the right place at the right time" for breeding. Rainey (337, 340) has made a notable contribution to the understanding of these migrations in terms of modern synoptic meteorology and he has emphasized the behavioral adaptation made by the insects, particularly in their exploitation of seasonal shifts in the intertropical convergence zone. Pedgley and Symmons (341) have discussed the 1967–1968 upsurge in locust populations in Africa, the Middle East, and India. They conclude that the principal cause was abnormal regional rainfall patterns.

In many cases, convergent wind flow has been found to play a part in generating swarms. Tropical cyclones (of size 10^5 km²) exhibiting convergence on both the macro- and the mesoscales have been suggested as a factor in concentrating locusts to high densities on at least six occasions. Observed displacements of swarms are shown in Fig. 75 (337, p. 21). The locusts move downwind but their average speed is somewhat less than that of the wind, probably because of intermittent settling of some of the insects on the ground. In this connection, most swarms fly by day and settle at night, although there is night flying if temperatures are sufficiently high. Figure 75 suggests, and it has been demonstrated in many instances, that it is possible to forecast the arrival of locust invasions from trajectory analysis; the June 1961 migration in India and Pakistan is an example (337, p. 30).

The mesoscale structure of swarms is also of interest. In many cases

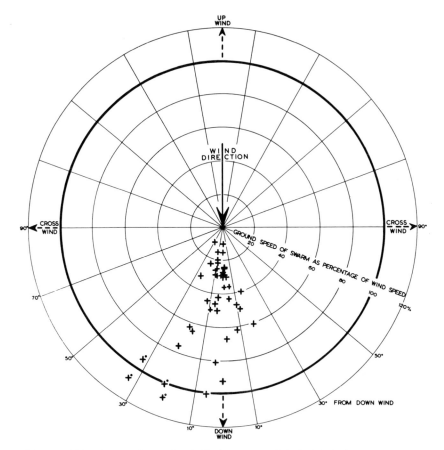

Fig. 75. Direction and speed of displacement of individual locust swarms in relation to the wind. The crosses with superscript dots indicate swarms flying up to more than 900 m above the ground. The data are from eastern Africa during the period 1951–1957 (337).

cohesion is maintained by the active behavior of the locusts themselves, turning back on arrival at the outer perimeter. Swarms may be disrupted by strong winds, to be reconstituted subsequently by localized convergence. In addition, Scorer (342) has suggested that convective cells may operate to maintain swarms rather than to disrupt them.

Locusts modify the local micrometeorology by adding metabolic heat and changing the radiation balance. Although Rainey (340) has suggested that the heating rate may amount to only about 0.1°C/3 hr, the effect, if any, on the behavior of the convective cell and on swarm motion remains to be elucidated.

Evidence is accumulating that distant migrations of many insect pests

are controlled by the wind; examples include invasions of the Oriental
army worm in China, the African army worm in Ethiopia, Kenya, and
Tanzania, the greasy cut worm in the Near East, and leaf hoppers in North
America (343). Huff (344) has examined 25 influxes of the potato leaf
hopper into Illinois during the period 1951–59. He finds that a persistent southerly flow occurred for at least the previous 36 hr, suggesting
a source region in the Mississippi delta. At the time of arrival in Illinois,
there were showers and a frontal passage, usually a cold front. Huff has
postulated that the leaf hoppers are washed out by rain but that the
sudden change in air-mass properties may also induce flight termination.
As an independent test, Huff searched the 1960 synoptic charts for cases
of southerly flows followed by a frontal passage. There were 7 such days,
and there were leaf-hopper increases on 6 of them; in the seventh case,
entomologists confirmed that source region populations were very low
at the time.

Birds, of course, have more control over their movements than do
insects. That migrations are affected by the weather is most readily
demonstrated by examining particular cases when there are large
anomalies in regional wind flows. Tropical birds have been carried northward into New England and Nova Scotia in the eyes of hurricanes. More
spectacular, however, have been the westward transatlantic lapwing
flights, which occurred in December 1927 and in January 1966. The lapwing is a European land bird that cannot alight on water and is rarely

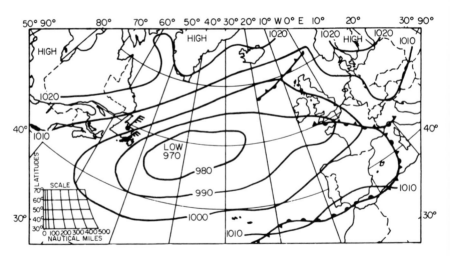

Fig. 76. Synoptic weather map for 0001 GMT, January 20, 1966. Easterly winds
extended from Ireland to Newfoundland.

seen in North America. Bagg (345) has documented the two migrations into the Canadian Atlantic provinces and New England, and he has described the associated synoptic weather. In both cases, there were unusual periods of easterly winds that swept the lapwing across the North Atlantic. The surface synoptic chart for 0000 GMT, January 20, 1966 (Fig. 76) is typical.

10.5. THE EFFECT OF CONTRAILS ON SURFACE TEMPERATURE: A SIMULATION MODEL

Aircraft condensation trails attenuate solar radiation. Reinking (346) speculates that this man-made form of weather modification may occasionally be helpful in reducing surface daytime air temperatures at times when heat waves or droughts would normally cause crop damage. Pearson (347) suggests that because the contrail-induced cirrus shield is usually a forerunner of an approaching cloud cover, natural processes have only been hastened by an hour or so. Nevertheless, when a storm system passes nearby or when there is a stationary front not too far distant, the possibility exists that the area of cirrus cloud is increased significantly by contrails (348). An experimental program designed to test the effect of condensation trails on surface temperature would be costly, requiring both aircraft and ground-based control facilities. The results would undoubtedly be inconclusive because of the difficulty in obtaining comparisons with an aircraft-free control area (rather similar uncertainties arise in cloud-seeding experiments). Nicodemus and McQuigg (349) have therefore developed a simulation model to examine the effect of contrails on surface temperature. The method is described in some detail as an example of the application of a Monte Carlo technique to a synoptic problem.

Weather observations were used for Columbia, Missouri for the months May–September (the growing season) during the years 1946–1965. The data included daily values of maximum and minimum temperatures, the percentage of possible sunshine, and upper air temperatures at the 350-, 300-, 250-, 200-, and 150-mb levels. A computer program was developed, which is illustrated schematically in Fig. 77.

In the first step, if surface maximum temperature is below 32°C (90°F) and/or if soil moisture is above 75% of field capacity, weather modification is assumed to have little effect on crop yield, and the day is rejected. The computer program can be adjusted readily if the limiting values of temperature and soil moisture are changed.

In the next step, the observed percentage possible sunshine is examined.

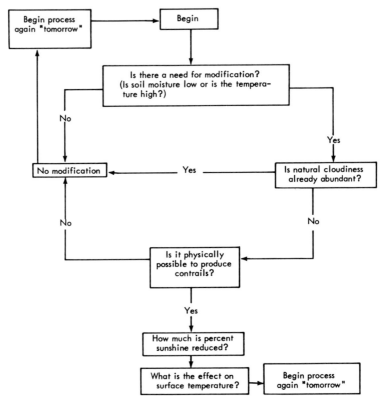

FIG. 77. Simulation model for examining the effect of aircraft contrails on surface temperature at Columbia, Missouri (349).

Because contrails are ineffective if skies are already cloudy, such days are not considered further.

The remaining cases are then classified according to whether an aircraft would or would not produce high cloud. This question can be answered only in probabilistic terms. The United States Air Weather Service has published a table of probabilities of contrail formation at various temperatures at each of the 350-, 300-, 250-, 200-, and 150-mb levels (350). For example, the likelihood of contrails at 350 mb is negligible at −39°C, 42% at −49°C, and 100% at −59°C. The computer generates a pseudorandom number, uniformly distributed between 0 and 100, and compares it with the probability of contrails at the temperature of each level in turn. If the random number is smaller than the probability for that level, the day is assumed to be favorable for the formation of contrails. Otherwise, the search continues through all levels, after which the day may be rejected.

The remaining sample consists of so-called "success" days, i.e., days (selected by the computer) when atmospheric conditions are assumed to be suitable for contrail production. The next step is to estimate the reduction in sunshine on "success" days. Again, a probabilistic approach is used, based on the equation

(10.2) $$S^* = S - S'$$

where

S^* = the modified percentage of sunshine due to high cloudiness,
S = the actual percentage of possible sunshine,
S' = a pseudorandom number drawn from a normally distributed population.

The S' distribution was chosen arbitrarily after study of particular days at Columbia, Missouri when there were varying amounts of naturally occurring cirrus, with no lower clouds. Because the quantity S was rarely less than 60% on such occasions, the mean and standard deviation of S' were selected so that S^* would hardly ever fall below 60%. Nicodemus and McQuigg recognize that the distribution of S' is uncertain, depending on the number of aircraft and the persistence of contrails. This does not detract from the power of the model, which can be modified when more information becomes available; in fact, three different distributions of S' were selected in order to test the sensitivity of the final results to gross changes in the S' distribution.

Finally, the increase in high cloudiness must be related to the decrease in surface air temperature. For a sample of days when cirrus alone occurred naturally, a linear regression was obtained:

(10.3) $$\Delta T = a + bS + e$$

where

ΔT = the rise in temperature from the early morning minimum to the afternoon maximum,
a, b = regression coefficients,
e = an error term normally distributed with a mean of zero.

Thus,

(10.4) $$\Delta T^* = a + bS^* + e$$
(10.5) $$T^*_{max} = T_{min} + \Delta T^*$$

On any particular day, a random value of e is generated by the computer

TABLE XXXII

AVERAGE DECREASES (IN °C) OF DAILY MAXIMUM AIR TEMPERATURES AT
COLUMBIA, MISSOURI IN THE GROWING SEASON, OBTAINED FROM THE
COMPUTER MODEL, ASSUMING THREE DIFFERENT PERCENTAGE
DECREASES IN SUNSHINE CAUSED BY AIRCRAFT CONTRAILS[a]

	15	25	35
Assumed average decrease in sunshine, %	15	25	35
Resulting average temperature decrease for all days	0.3	0.4	0.5
Resulting average temperature decrease for "success" days	1.7	2.2	2.6
Resulting average temperature decrease for all days when modification is needed	0.9	1.2	1.4

[a] Nicodemus and McQuigg (349).

from a table of possible values. The modified maximum temperature T^*_{max} is then determined from Eqs. (10.4) and (10.5) and compared with the actual T_{max}. The assumption is made that contrails do not affect the minimum temperature of the previous night. Over many trials, the effect of e becomes negligible. However, Nicodemus and McQuigg believe it to be important to retain the random term in a simulation study.

Table XXXII displays the computer results obtained from 20 years of data at Columbia, Missouri. Three different distributions of S' in Eq. (10.2) were used, based on average decreases in percentage of sunshine of 15, 25, and 35% due to contrail modification. Because contrails were assumed in the model to have no effect or to be of no agricultural value on many days, the seasonal reduction in mean maximum temperature was small, less than 0.5°C. On "success" days, the average decrease ranged from 1.7 to 2.6°C; including days when modification was needed but not achieved, maximum temperatures were reduced on the average by 0.9–1.4°C.

Simulation studies require physical insight into the problem. However, they may be valuable tools in biometeorological investigations.

10.6. SYNOPTIC MAXIMIZATION TECHNIQUES

On the synoptic scale, hydrometeorologists frequently assume that there is a physical upper limit to the amount of precipitation that can fall in a storm. This limit is rarely if ever reached but it is a constraint in the design of flood-control systems. The method has been described by Bruce and Clark (351, p. 230) and by Pullen (352).

From historical records, several heavy-rainfall cases are selected. Isopleths of rainfall are drawn (isohyetal maps), and the synoptic surface and upper air conditions are studied. Each storm is then "maximized," using three adjustments:

(a) The central isohyet is shifted slightly to correspond more closely to the center of the basin under investigation.

(b) If the isohyets are elliptical rather than circular, they may be rotated to correspond to the major axis of the basin. The degree of rotation and of translation requires a judgement based on synoptic experience with storm tracks and orographic influences. This can only be done by a skilled meteorologist.

(c) The amount of precipitable water in a vertical column is increased by assuming that the air is saturated, that the temperature lapse rate is pseudoadiabatic, and that the surface dewpoint, averaged over 12 hr, is a climatological maximum for the basin.

An exchange of letters in the *Meteorological Magazine,* 1967, pp. 348–351, concerns the relative merits of storm maximization and extreme value (Section 6.6) methods in hydrology. Storm maximization estimates yield no indication of the probability or return period of the limiting critical event; the statistical approach, on the other hand, assigns a probability that a given rainfall at a point will not be exceeded but gives no indication of a physical upper limit or of the areal average. Singleton (353) suggests that if loss of life is a consequence of the critical flood, the storm maximization method is preferred. Yevjevich (354), however, believes that an unreasonable economic burden is placed on the engineer if a return period cannot be specified.

Lockwood (355) has attempted to assign probabilities in a storm maximization study. Suppose that A_1 is a large, climatically homogeneous area with a satisfactory coverage of rain gauges, and that A_2 is the average area enclosed by the X-cm isohyet, a rainfall amount that occurs on the average only once in t years. Then the probability of more than X cm occurring at any point within the region during t years is A_2/A_1. Assuming independence of space and time, the return period for X cm is therefore $(A_2/A_1)(1/t)$. Lockwood compares his formula with that obtained by the Gumbel method, using data from Malaya, and finds reasonable agreement.

Precipitation maximization methods are used also in glaciology to estimate maximum growth rates in prehistoric times and to examine present-day accumulations. An example of nonlinear interactions has been given by Lettau (356) for the Byrd station in Antarctica. He found that the saturated mixing ratios derived from mean monthly vertical temperature profiles underestimated the observed snow accumulations;

there was evidence that moisture advection during brief intrusions of warm air was a major contributor to the snow pack.

10.7. ECOLOGICAL METEOROTROPISMS

The 1952 weekly death rates from respiratory and heart diseases in London, England display a very large peak in early December (see Fig. 3). This ecological meteorotropism is so marked that inductive reasoning connecting death rates with atmospheric conditions is convincing. Physiological studies confirm that high concentrations of SO_2 are harmful in the presence of fog. In other instances, a number of weather elements may all show concurrent peaks or rapid changes, and there is no way of knowing which, if any, are causing stress. To overcome this problem, Grosswetterlagen and frontal passages have been used frequently as meteorotropic indices.

Driscoll and Landsberg (357) have examined the effect of a cold front on mortality. Although their results are negative, the methodology is important. The usual approach is to examine data from a fixed location over a long period of time, attempting to correlate peaks in mortality rates with frontal positions or abrupt temperature changes. Instead, Driscoll and Landsberg followed the movement of a sharp cold front across the United States during the period October 24–29, 1963. The antecedent weather was warm and relatively settled for 1–2 weeks prior to the arrival of the front.

Mortality figures were obtained for deaths due to vascular lesions affecting the central nervous system, commonly referred to as "strokes." Some of the results are given in Table XXXIII. Using the frontal passage as an origin of time, death rates show little change in consecutive 10-day

TABLE XXXIII

MEAN NUMBER OF DEATHS PER DAY PER ONE MILLION POPULATION DUE TO VASCULAR LESIONS AFFECTING THE CENTRAL NERVOUS SYSTEM FOR PERIODS 10 DAYS PRIOR TO, AND 10 DAYS AFTER FRONTAL PASSAGE[a]

State	Prior	After	State	Prior	After
Arkansas	3.4	3.8	Missouri	3.3	3.4
Georgia	3.6	3.9	New Jersey	2.3	2.2
Illinois	2.4	2.6	Ohio	2.6	2.3
Indiana	2.7	3.4	Tennessee	3.2	3.8
Kentucky	3.9	3.7	Virginia	2.8	2.3
Mississippi	3.7	2.8	West Virginia	3.3	2.8

[a] Driscoll and Landsberg (357).

periods. It is concluded that for this particular synoptic case, there is no evidence of a meteorotropism.

Driscoll and Landsberg's study is almost unique because rarely does a negative result appear in print. Many of the reported meteorotropisms are based on rather slender statistical evidence, and Sargent (358) in fact suggests that until there is confirmation from controlled laboratory experiments, the doctrine "is no more secure than a house of cards."

Physiological links have been demonstrated for the stresses created by ragweed pollen, oxidant pollution, and heat waves. Synoptic studies can be very fruitful in such cases and may form the basis for a warning or alert service. For example, death rates have been studied by Bridger and Helfland (118) during a heat wave in Illinois, July 10–14, 1966. Death rates for the summer of 1965, considered to be a more normal year, were used as a control. There was no indication of a meteorotropism in either year among people from 45 to 64 years of age for deaths from cardiovascular disease. Among the elderly, however, a significant increase in deaths occurred during the heat wave. Bridger and Helfland therefore recommend that routine weather forecasts should include warnings of heat-stress conditions.

The winter of 1962–1963, particularly February, was extremely cold in Britain. There was also a week of severe smog in early December. Carne (359) has discussed the weekly totals of respiratory disease onsets in London and Sheffield for the winters of 1962–1963 and 1963–1964. At first glance, the London values strongly support the existence of meteorotropisms. There was a sharp peak in respiratory onsets during the December smog episode but an even larger increase in February; neither of these was repeated in 1963–1964. The December peak was missing in Sheffield, a fact that might be explained by lower levels of pollution. However, there was also no increase in February, casting doubt on the hypothesis that the cold wave was a specific causative agent for respiratory disease. In fact, there were slightly more respiratory onsets in 1963–1964 than in 1962–1963 in Sheffield. These data illustrate the dangers of drawing inferences about health from meteorological spells. McCarroll (117) agrees; he notes that in New York City there are days when mortality rates are relatively high but pollution levels are low; there are also days when mortality rates are low but air quality is poor. However, "the sharpest peaks in mortality are most consistently associated with sharp rises in air pollution occurring almost simultaneously," which tempts the investigator to believe that a tenuous connection does in fact exist.

11/SEASONAL RELATIONSHIPS

11.1. GENERAL REFLECTIONS

This chapter is concerned with time periods ranging from a week to a year. Although the meteorologist feels most secure with synoptic data, there are many physical and biological processes that should or must be studied over longer time scales. The water budget of a lake or watershed frequently can be estimated only over a minimum period of a month. The growth of a crop continues throughout a summer, and the yield is dependent on the seasonal integration of day-to-day weather.

Statistical methods frequently fail on these longer time scales. For a crop that is planted in spring and harvested in autumn, the life span is not a stationary time series. Not only does a plant respond to environmental stresses differently at different stages of its life cycle, but the environmental stresses themselves also exhibit seasonal trends (in number of hours of daylight, air temperature, soil moisture availability, etc.).

Turning now to another question, physical and statistical models are not yet capable of describing meteorological persistence in any realistic way. A specific weather type may last a week or, in the case of the monsoon, a season. The change to another macroscale pattern is sometimes dramatically abrupt, as when a temperate-zone "blocking" anticyclone develops. Termed *intermittency*, this "square-wave" phenomenon occurs on all time scales. Garnier (360) has noted that in Nigeria, for example, regular climatic averages mask the basically intermittent character of the weather. The annual cycles of most of the meteorological elements in any one year in that country resemble saw-tooth rather than smooth sinusoidal curves. The surface discontinuity between moist and dry air moves north and south with the season, but in spurts. Garnier believes that four weather patterns can be identified:

Zone A: North of the surface discontinuity: dry and sunny.

Zone B: Immediately south of the surface discontinuity: daytime scattered cumulus clouds and high humidities but dry air aloft.

Zone C: Farther south: variable cloudiness, high humidities, and showers.

Zone D: Farther south again: considerable low cloudiness and high humidities, no significant precipitation.

The basic intermittency in the motion of the surface discontinuity must be recognized if Nigerian climate is to be understood. There is merit, therefore, in describing the local climate in that country by the monthly frequencies of each of the four types of weather patterns.

11.2. EVAPOTRANSPIRATION ESTIMATES

Evapotranspiration (ET) is the rate of water loss per unit area of vegetated surface, by transpiration from the leaves and by evaporation from the underlying surface. *Potential evapotranspiration* (PET) is the rate of water loss that would occur if soil moisture over a large area were not a limiting factor. *Local potential evapotranspiration* (LPET) is the rate of water loss from a moist surface so small that the transpired water vapor has no appreciable influence on the moisture content of the air passing over the "oasis." In the case of bare ground or open water, the appropriate terms are *evaporation* E, *potential evaporation* PE, and *local potential evaporation* LPE. A pan of water in the desert yields values of LPE. In general,

$$\text{LPET} \geq \text{PET} \geq \text{ET} \quad \text{and} \quad \text{LPE} \geq \text{PE} \geq \text{E}$$

A long-time goal of hydrologists, agronomists, geographers, and others has been to estimate E, ET, PE, and PET from standard climatic data and/or from evaporation pan measurements of LPE, averaged over a day or a month. It is clearly desirable to determine the water losses from the surface of the earth. In addition, the ratio ET/PET or the difference (PET − ET) is an index of soil moisture stress, which in turn is correlated with plant yields.

The quantities E and ET can be measured directly over bare soil and over vegetation using lysimeters (10, p. 32) but there are few of these instruments throughout the world. Hourly estimates can also be obtained by micrometeorological methods and summed to give a monthly total; here again, however, special instruments are required. Fnally, area monthly values of E and ET may be inferred from the water budget equation but it is desirable to obtain independent estimates.

The quantities PE and PET cannot be measured except during and shortly after a period of wet weather, when the "potential" is realized. In the simplest case, flow over an ocean or large lake, it might be thought that PE and E would be equal. However, Bean and Florey (361) have estimated that the evaporation from Lake Hefner, Oklahoma was reduced by 58% when a monomolecular film was applied. Many bodies of water contain oil slicks or other impurities which increase the resistance to evaporation through the viscous sublayer. It cannot therefore be assumed that PE = E. For a pan of water too, the actual evaporation is less than LPE unless distilled water is used and changed regularly. Another complexity is that a deep-rooted plant may transpire at the potential rate at times when seedlings are undergoing moisture stress.

Consider next a very large vegetated area under moisture stress and suppose that irrigation is begun. The characteristics of the surface boundary layer begin to change. At first, the evapotranspiration is very great because of the large vapor pressure difference existing between the interface and the air. Later, however, the rate decreases as the air becomes more humid. There are in fact two limiting cases for which ET = 0:

(a) The soil is bone dry.
(b) The air is saturated, with zero or negative temperature lapse rate.

Note that if the surface is warmer than the air, evaporation continues, even when the air is saturated, although at a reduced rate (see Table XXVII). The air does not usually become saturated, however. Even over a large lake or an ocean, the relative humidity at a height of a few meters is hardly ever 100% unless the flow is over increasingly cooler water. Sea fog forms only during these latter conditions.

These comments suggest that the definitions of PET and PE are not very precise. It is not clear whether "potential" conditions exist shortly after a large area is irrigated or at some later time when equilibrium is achieved, although in arid regions, this latter condition may never be reached. Fritschen and Nixon (362) have emphasized that the atmosphere at heights of a few meters is not modified substantially by irrigation in the San Joaquin Valley of California. There are two reasons for this. In the first place, there is an almost unlimited supply of dry air in the surrounding mountains; drainage winds create a daily exchange. Second, because irrigation is not continuous but is rotated throughout the valley, there are always dry areas at any given time.

Daily observations of LPE from evaporation pans are available in many parts of the world. In Ottawa during summer (363, p. 16), for example, the water loss from a pan requires about 50 kcal of energy whereas the net radiation totals only 40 kcal; advection must therefore supply an additional amount.* The diameter of the pan and the rim-height above ground affect the internal boundary layer that develops; in addition, some pans in tropical countries are protected from insects by screening, which changes the character of the turbulence at the water surface. Each type must therefore be "calibrated" empirically if extrapolation to PE or PET is to be attempted; a useful summary of this question is included in a WMO Technical Note (364). Only qualitative results can be expected, although these may be of value in ranking a series of summers. The same comments apply to atmometers, which are devices for measuring the evaporation from a porous disk or filter paper (364, p. 8). Here again, extrapolation to PET or ET cannot be made on physical grounds, although some investigators may determine empirical relations of local value.

Yu and Brutsaert (365) comment that "in spite of the vast amount of experimental data, there is no evidence in the literature that the phenomenon of evaporation from pans is completely understood." In addition to the microscale uncertainties at the pan itself, there is the question of exposure, which is just as important a consideration in evaporation as in rainfall measurements (see Section 3.6); care must be taken not to overexpose a pan. The further problems of size and color have been examined by Yu and Brutsaert, who undertook a comparison of square pans of standard depth 2.5 cm (1 in.), of three lengths, 0.3, 1.2, and 2.4 m (1, 4, and 8 ft), and of two colors, black and white. For the three black pans, the evaporation rate increased with size but the opposite trend existed in the case of the white pans. The correlation was not high between measured evaporation and solar radiation; however, there was a significant improvement when the difference in evaporation between black and white pans of the same size was correlated with solar radiation.

Following this introduction, it is now appropriate to examine some of the "standard" methods for estimating monthly values of PE and PET. Many of the approaches have been influenced by the desire to utilize standard climatological data. In the case of evaporation pans, for example, Yu and Brutsaert (365) emphasize that the pan is a very old tool, and it will "continue to be the only cheap and generally accepted

---

* In arid regions, the latent heat flux over pans may be twice as large as the net radiation.

instrument available to hydrologists and engineers" for many years to come, especially in remote regions.

The *Dalton equation* is sometimes used to estimate lake evaporation (364, p. 80):

(11.1) $$E = f(\bar{u})(e_s - e_a)$$

where

$f(\bar{u})$ = a function of mean wind speed $\bar{u}$ and is determined empirically,
e_s = the saturated vapor pressure at the temperature of the water surface,
e_a = the vapor pressure of the air at some standard level.

The function $f(\bar{u})$ is often represented by the expression $c_1(1 + c_2\bar{u})$, where c_1 and c_2 are empirical constants, the value of c_1 decreasing slightly as the size of the lake or reservoir increases. Equation (11.1) has been fitted to both daily and monthly mean winds and vapor pressures; the values of the constants vary accordingly. Reviews have been given by Bruce and Clark (351, pp. 58, 98) and by a WMO Working Group (364, p. 80).

Harbeck has found that for various sizes of reservoirs (366):

(11.2) $$E = 0.291A^{-0.5}\bar{u}_2(e_0 - e_2)$$

where

E = the reservoir evaporation in centimeters per day,
A = the reservoir area in square meters,
$\bar{u}_2$ = the mean wind speed at 2 m, in meters per second,
e_0 = the saturated vapor pressure at lake temperature, in millibars,
e_2 = the vapor pressure at 2 m, in millibars.

The quantities $\bar{u}_2$, e_0, and e_2 are all measured in the middle of the reservoir. The relation is empirical but the fact that the area of the reservoir is related inversely to evaporation suggests that the water loss is intermediate between LPE and PE.

For hourly values of the variables during steady-state conditions, Eq. (11.1) has some theoretical justification, being a form of the Thornthwaite–Holzman equation (10, p. 93). Even then, however, its validity is limited to near-adiabatic conditions. When daily or monthly mean values are employed, the time-averaging problem discussed in Section 2.1 becomes important; winds are lighter and evaporation is less at night than in the daytime. Equation ((11.1) has been applied also in the determination of qualitative 24-hr estimates of evaporation over oceans; in this case the assumption of steady-state adiabatic conditions persisting for 24 hr is sometimes realized.

Another empirical approach suggested by Thornthwaite (367) has been used for many years:

(11.3) $$PET = (h/12)(n/30)F(\bar{T})$$

where

PET = the potential evapotranspiration measured in cm/month,
h = the duration of daylight in hours,
n = the number of days in the month,
$F(\bar{T})$ = a complicated function of the mean monthly temperature, $\bar{T}$, and of a weighted annual mean temperature.

For $\bar{T} < 0°C$, it is assumed that PET = 0. For $\bar{T} = 26.5°C$, PET = 13.5 cm/month in all climatic regions. Estimates of PET are based therefore only on astronomical variables and on mean temperatures. The factors $(h/12)$ and $(n/30)$ normalize the equation for season, latitude, and number of days in a month. Thornthwaite (368) included data from Florida, New Mexico, Ohio, Nebraska, Colorado, and Wyoming in his paper and proposed that

(11.4) $$PET = c_1 \bar{T}^{c_2}$$

where c_1, c_2 are constants for each site. According to Thornthwaite, Eq. (11.4) is "far from satisfactory" and is "completely lacking in mathematical elegance."

A fundamental question that must be asked is why there should be a useful relation at all between PET and $\bar{T}$. Pelton et al. (369) seem to have the correct answer. For monthly means and for moist surfaces, both $\bar{T}$ and PET are correlated with monthly mean net radiation; thus there must be a correlation between PET and $\bar{T}$. These authors suggest that any variable that is highly correlated with monthly mean net radiation (such as soil temperature) would be a useful predictor of PET even though there is no physical relation.

Högström (370) has calculated hourly values of ET from vertical gradients of temperature and humidity for a period of 29 months at Ugerup in southern Sweden. These he has summed by months and compared with values of PET obtained from Eq. (11.3). The ratio of PET/ET is given in Fig. 78, and a curve has been drawn by hand that seems to fit the trend. However, Högström notes that the soil is usually wet in September and October; the ratio should then be close to unity instead of near the value 2. At Ugerup, at least, it seems that Thornthwaite's method is not satisfactory. Högström has also compared pan evaporation with values of ET: the ratio of LPE to ET averaged 1.37 for the frost-free months but the individual monthly values were scattered with no discernible trends.

PET/ET

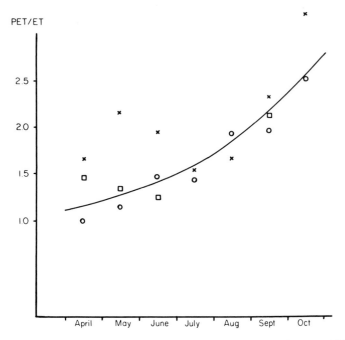

FIG. 78. The monthly ratio PET/ET at Ugerup, Sweden; (×) 1961; (○) 1962; (□) 1963. PET is estimated by Thornthwaite's method and ET is obtained by summing hourly values computed by micrometeorological methods (370).

Penman (371) has suggested an equation of the following form:

(11.5) $$\text{PET} = F(Q_N, \bar{u}, e_s, e, de_s/dT)$$

where

Q_N = net radiation,
$\bar{u}$ = mean wind speed,
e_s, e, T = the saturated and actual vapor pressures and temperature at screen level.

The model has been used with daily, weekly, or monthly mean values of the variables. The inclusion of net radiation is physically satisfying and restricts the magnitude of the errors that can be made. Högström has computed Penman's PET for monthly intervals at Ugerup, Sweden. A comparison with the values of ET obtained by summing micrometeorological estimates is given in Fig. 79. During the spring and summer, the ratio PET/ET is greater than unity, the very high spring value being due to low precipitation and generally dry conditions. The winter value of the ratio is less than one, which is impossible by definition. Högström

PET/ET

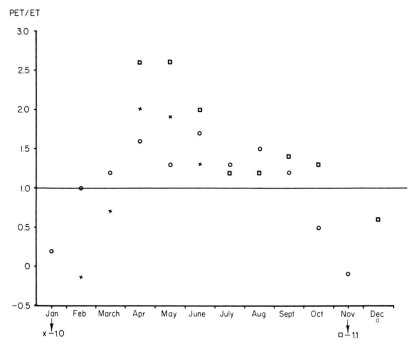

FIG. 79. The monthly ratio PET/ET at Ugerup, Sweden: (□) 1961; (○) 1962; (×) 1963. PET is estimated by Penman's method and ET is obtained as in Fig. 78 (370).

concludes that the problem is twofold: (a) the neglect of soil heat storage in Eq. (11.5), and (b) the use of a monthly averaging time for calculations of PET. In winter, warm moist periods with low evaporation alternate with cold dry periods when moisture flux is high; monthly averages of the meteorological elements than underestimate PET.

For irrigated pasture land in South Australia, Holmes and Watson (372) have compared the values of PET given by (a) a number of empirical formulas, (b) a lysimeter, and (c) the monthly water budget equation. Significant regression equations were found in all cases, with correlation coefficients of about 0.8. However, a direct regression between PET and Q_n gave a correlation coefficient of the same magnitude, which supports Penman's view that net radiation is the most important measurement to be made in water budget studies.

Whereas Thornthwaite and Penman have used observations at screen level only, Tanner and Fuchs (373) have developed a model that includes values of temperature and vapor pressure at the interface itself. The heat and mass transfer resistances r_H and r_E are introduced [see Eq. (8.28)]:

(11.6) $Q_H = \rho c_p (T_0 - T_z)/r_H$

(11.7) $Q_E = L(\rho \epsilon / P)(e_0 - e_z) r_E$

where

$\epsilon = 0.622$, the ratio of mole weights of water vapor and air,
P = the barometric pressure,
T_0, e_0, T_z, e_z = the temperature and vapor pressure at the interface and level z, respectively.

Assuming $r_H = r_E = r$, it can be shown that,

(11.8) $ET = [s/(s + \gamma)][Q_N - Q_G + (\rho c_p/rs)\{(e_z{}^* - e_z) - (e_0{}^* - e_0)\}]$

where

s = the slope of the curve of the saturation water vapor plotted against temperature [as in Penman's Eq. (11.5)],
γ = the psychrometric constant, $c_p P/\epsilon L$,
$e_z{}^*, e_0{}^*$ = the saturated vapor pressures corresponding to T_z and T_0.

Equation (11.8) applies in the case of a surface that is only partially wet, such as a drying soil. If the interface is in fact moist, $ET = PET$ and $e_0{}^* = e_0$. The equation then simplifies to

(11.9) $PET = [s/(s + \gamma)][Q_N - Q_G + (\rho c_p/rs)(e_z{}^* - e_z)]$

If the air at height z is also saturated,

(11.10) $PET = [s/(s + \gamma)] (Q_N - Q_G)$

Equations (11.9) and (11.10) cannot, however, be used to calculate PET from measurements taken during periods of moisture stress. If the area were irrigated at that time, leaf temperature would decrease, and the values of both Q_N and Q_G would change.

There remains the difficulty of estimating r_H or r_E, a requirement for solving Eqs. (11.8) and (11.9). Prouitt and Lourence (374) have derived an empirical form, based on the wind speed at a height of 1 m at Davis, California, while Hunt *et al.* (375) have measured the resistance of an individual sunflower leaf directly with a diffusion porometer. Nevertheless, a "universal" model has not yet been established. Sziecz and Long (376) have, in fact, suggested that experimental values of ET be used to estimate the resistance. They have described methods of doing this and of separating the resistance into its leaf-surface and atmospheric turbulent components [see Eq. (8.38)]. They recommend the use of the leaf-surface resistance as an index of moisture stress.

A useful result has been given by Priestley (377, p. 116) for a saturated surface.

(11.11)
$$Q_\text{E}/Q_\text{H} = \frac{L}{c_\text{p}} \frac{(\partial q_\text{s})}{\partial T}$$

where q_s is the saturated specific humidity at surface temperature T. The relation may be tabulated from psychrometric tables, and its form is shown graphically in Fig. 80. As the temperature increases, more and more of the available energy is used for evapotranspiration. This result has been exploited by Munn and Truhlar (378), who have *defined* a condition of PET as one for which Eq. (11.11) applies at the interface.

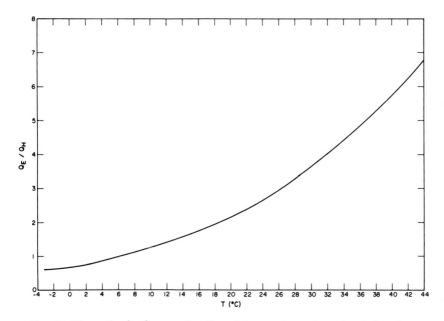

FIG. 80. The ratio Q_E/Q_H as a function of temperature at a saturated surface.

Let the daytime energy balance of a short grass surface (see Eq. 8.19) during a period of soil moisture stress be,

(11.12)
$$Q_{\text{N}_1} - Q_{\text{G}_1} = Q_{\text{H}_1} + Q_{\text{E}_1}$$

This is changed by irrigation to

(11.13)
$$Q_{\text{N}_2} - Q_{\text{G}_2} = Q_{\text{H}_2} + Q_{\text{E}_2}$$

where suffixes 1 and 2 refer to conditions before and after irrigation, and $Q_{\text{E}_2}/Q_{\text{H}_2}$ is a constant for given surface temperature T_2.

Let

$$Q_{\text{E}_2}/Q_{\text{H}_2} = 1/B$$

TABLE XXXIV

ESTIMATES[a] OF Q_{E_2} (LY/HR) FOR $T_1 = 30°C$, $Q_{N_1} = 51$ LY/HR, AND $Q_{G_1} = 0.1\ Q_{N_1}$[a,b,c]

Q_{G_2}/Q_{G_1}	T_2	
	25°C	20°C
1.5	33.8	33.2
2.0	32.1	31.6
2.5	30.1	29.8

[a] Munn and Truhlar (378).

[b] See Eqs. (11.12) and (11.13) for notation.

[c] The values of Q_{N_2} are obtained from the assumed values of T_2 and of the radiation components.

where B is the Bowen ratio. Then from Eq. (11.13),

$$(11.14) \qquad Q_{E_2} = (Q_{N_2} - Q_{G_2})/(1 + B)$$

Because the surface temperature has decreased after irrigation, the long-wave upward radiation component has decreased and $Q_{N_2} > Q_{N_1}$; in addition, $Q_{G_2} > Q_{G_1}$ because the soil is moister. Munn and Truhlar have examined particular cases, determining the percentage change in Q_E associated with various estimates of the relative magnitudes of the other terms in Eqs. (11.12) and (11.13). One example is typcial, in which $T_1 = 30°C$, $Q_{N_1} = 51$ ly/hr, the emissivity of the surface is assumed to be unity, $Q_{G_1} = 0.1\ Q_{N_1}$, and reasonable values of the short-wave and

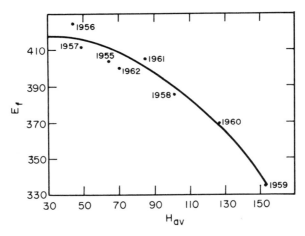

FIG. 81. Relationship between evaporation from a forest E_f and the average ground-water level H_{av} for May to September in the Taezhnii area of the U.S.S.R. (379).

downward long-wave components are chosen. Estimates of Q_{E2} are given in Table XXXIV for two selected values of T_2 and three values of the ratio Q_{G2}/Q_{G1}. The resulting values of Q_{E2} range from 33.8 to 29.8 ly/hr, a difference of less than 15%. A variation of this magnitude is to be expected because of the wide range of possible initial soil moisture deficiencies. When the investigator has a knowledge of prevailing local conditions, he should be able to estimate PET to within $\pm 5\%$. Fritschen and Nixon (362) reach about the same conclusion, using a Penman-type approach, substituting data obtained from desert and irrigated areas in the San Joaquin Valley of California.

Finally, another seasonal index of ET is the average ground-water level. Each soil type and vegetation cover must be considered separately. Nevertheless, useful prediction curves can sometimes be obtained. Figure 81 was prepared by Fedorov (379) for the forested catchment of the Taezhnii ravine in the U.S.S.R.

11.3. SOIL MOISTURE BUDGETS

There are many who feel that soil moisture is the best environmental predictor of crop yield. Before considering the question of productivity in Section 11.5, therefore, recent studies of soil moisture budgets will be described. The word "budget" in this context implies the partition of soil moisture by layers and by time of year; the available water in the root zone is naturally of most importance, and this layer varies with the type and development stage of vegetation.

A plant is the ultimate judge of whether it is suffering from moisture stress. Unless the wilting point has been reached, however, there is no easy way of obtaining quantitative measurements of intermittent stress through the growing season for correlation with yield. Analysis of the week-to-week changes in the soil moisture profile is a useful alternative, but soil moisture probes are not yet generally available at all agricultural experimental stations for each different field crop. Baier and Robertson (380) have therefore attempted to model the partition of available water, using only standard meteorological measurements as predictors. Assuming that the model can be verified with actual soil moisture data, in a few cases, the historical records of crop yields may then be correlated with moisture stress.

Baier and Robertson undertook comparative experiments at Ottawa, Canada, where daily soil moisture measurements were available at depths of 0.25, 1, 3, and 6 cm, using Colman electrical resistance blocks.

Wheat was chosen for study and the growing season was divided into six time periods, separated by the following indicators of development: planting, emergence, jointing, heading, milk stage, soft dough stage, and hard dough stage. A lower limiting depth of 6.6 cm was selected; in this layer the readily available moisture was determined to be 1 cm at field capacity. The soil was then subdivided into six zones. These were chosen not directly as fractions of the total depth but according to water-holding capacities of 0.05, 0.08, 0.12, 0.25, 0.25, and 0.25 cm of water from the surface downward.

A computer program was developed to produce daily estimates of ET and of PET. Actual evapotranspiration was obtained as the difference between rainfall and the daily integrated soil moisture changes over all zones. Any excess of ET over PET was assumed to be runoff. Although PET was estimated with an atmometer and is therefore an uncertain quantity in the subsequent analysis, the general methodology is not affected.

A prediction model for ET was developed, which has the general form

$$(11.15) \qquad \mathrm{ET}_i = \sum_{j=1}^{6} \left[k_j \frac{S'_{j(i-1)}}{S_j} Z_j \, \mathrm{PET}_i \, e^{-w(\mathrm{PET}_i - \overline{\mathrm{PET}})} \right]$$

where

ET_i = actual evapotranspiration for day i, beginning in the morning and ending in the morning of day $i + 1$, with the summation carried out from zone $j = 1$ to zone $j = 6$,

k_j = coefficient accounting for soil and plant characteristics in the jth zone,

$S'_{j(i-1)}$ = available soil moisture in the jth zone at the end of day $i - 1$, that is, at the morning observation of day i,

S_j = maximum water-holding capacity,

Z_j = adjustment factor for different types of soil dryness curves,

PET_i = potential evapotranspiration for day i,

w = adjustment factor accounting for effects of varying PET rates on ET/PET (see below),

$\overline{\mathrm{PET}}$ = average PET for month or season.

The model contains the following assumptions:

(a) Water is withdrawn simultaneously from different depths down to the lowest level of root penetration. It is assumed that if the jth zone is at maximum water-holding capacity, ET = PET for that layer.

(b) If the jth zone is at less than 100% of its water-holding capacity,

ET < PET for that zone, although other zones may be transpiring moisture at the PET rate. This is a meaningful extension of Thornthwaite's concept of PET.

(c) An assumption is required concerning the relation between the ratio ET/PET and the percentage of possible soil moisture. Several suggested curves are given in Fig. 82, but the straight line C was selected. A subsequent study of short grass in Ottawa by Baier (381) yielded the results displayed in Fig. 83, suggesting that Type C was indeed a good choice. However, the parameter Z_j is retained in Eq. (11.15) to provide more generality; when the most appropriate drying curve is determined experimentally for a particular soil, the relation may be used in Eq. (11.15).

(d) The coefficients k_j are introduced to account for the fact that the roots are not distributed uniformly throughout the soil. If no roots have reached the bottom zone, moisture from that layer makes no contribution to ET, and $k_6 = 0$. The coefficient k_j therefore represents the fraction of roots in the jth zone, the sum of the k_j's totaling unity. The values of the j's must change during the season and can only be determined from crop-rooting patterns.

(e) The exponential term on the extreme right-hand side of the equation is included because experimental evidence indicates that a plant absorbs more moisture from the lower zones when the upper zones are dry, than when they are wet. The exponential term distributes part of the upper-zone coefficients into the lower zones when the upper layers are dry.

The usefulness of this method is considered in Section 11.5.

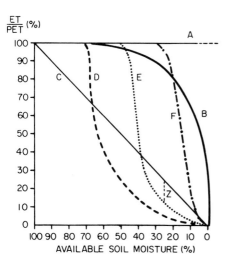

FIG. 82. Five types of assumed relations (B–F) between soil moisture availability and the ET/PET ratio (381).

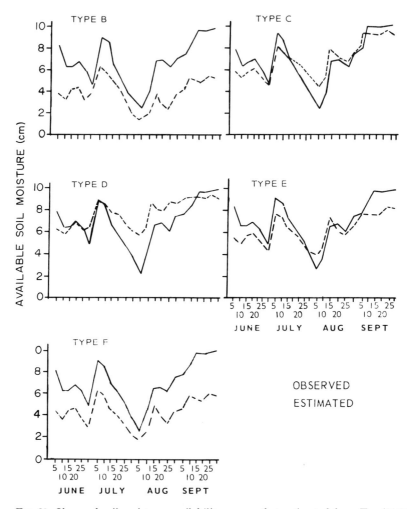

Fig. 83. Observed soil moisture availability versus that estimated from Eq. (11.15), using each of the five drying curves B–F in Fig. 82; the data are for a soil covered with short grass at Ottawa, Canada (381).

A simpler but less complete representation is the soil moisture deficiency integrated over all layers of interest. The rate of water removal from the soil by plants is often assumed to be proportional to the amount remaining in the soil, which leads to an exponential relation. Moisture deficiency has been estimated experimentally by Smith (382):

$$(11.16) \qquad\qquad D = AZ(Y - S)$$

where D is soil moisture deficit at soil moisture S (percentage of dry weight), A is apparent specific gravity (volume weight), Z is rooting depth, and Y is field capacity (percentage of dry weight). Smith obtained a significant correlation between the logarithm of the monthly number of diseased cacao pods in Trinidad and the value of D for the fourth preceding month.

Another index is the number of days with moisture stress during the growing season. Dale and Shaw (383) found a significant correlation between this quantity and the corn yields at Ames, Iowa. They have also published a 30-year climatology of moisture-stress days at Ames (384).

11.4. DROUGHT

In regions where most of the summer rainfall comes from thunderstorms, a community may experience local drought while adjacent areas have ample moisture. Of more general concern, however, is widespread drought caused by large-scale weather anomalies or by human intervention (improper agricultural practices or poor watershed management).

In Saskatchewan, serious drought usually results from two or more consecutive dry years (385); the time lag is much less in regions where soil moisture supplies are always marginal. This suggests as a first principle that time constants for the hydrologic cycle should be estimated for each area, soil type, and vegetative cover. Palmer (386) discussed the abnormally dry weather of 1961–1966 in the Northeastern United States. Noting that almost half of the precipitation becomes runoff during normal and above-normal years in that area, he suggests that when rainfall is deficient, agricultural yields may remain excellent despite major water shortages in lakes and reservoirs. Precipitation only 80% of normal in New York State is equivalent to average conditions in eastern Nebraska and western Iowa, an agriculturally productive region. This implies that there is more than one kind of drought, and that the definition varies according to the particular interest of the user. Even the agronomists and the foresters may not agree. There may be a serious water deficiency for a short-rooted cultivated crop at a time when adjacent forests are not experiencing moisture stress. Subrahmanyan (387, p. 37) recommends that at the very least, it should be possible to adopt a standard spelling and pronunciation of the word "drought."

A simple but not entirely satisfactory way of describing drought is in terms of precipitation deficiency. There are many such definitions, a common one being the occurrence of 15 consecutive days, each with no more than a trace of rainfall (387, p. 16); alternatively, a drought can be defined as a period of prescribed length, such as a season or a year, during which time the precipitation is less than a given fraction of the average value for the location, a normalization technique which assists in regional comparisons.

A drought can result not only from a precipitation deficiency, but also from excessive evapotranspiration; in foggy coastal areas, for example, a dry spell is not particularly harmful to vegetation, although there may be serious hydrological consequences.

Palmer (388) has developed a *meteorological drought index* PI (the Palmer index) and a *crop moisture index* CMI, both of which are calculated weekly during the growing season and disseminated widely in the United States. Descriptive terms associated with ranges of values of PI are given in Table XXXV (389). The key phrase is "relative to the

TABLE XXXV

THE PALMER METEOROLOGICAL DROUGHT INDEX[a]

Value	Class (relative to the particular location)
≥ 4.00	Extremely wet
3.00–3.99	Very wet
2.00–2.99	Moderately wet
1.00–1.99	Slightly wet
0.50–0.99	Incipient wet spell
0.49 to −0.49	Near normal
−0.50 to −0.99	Incipient drought
−1.00 to −1.99	Mild drought
−2.00 to −2.99	Moderate drought
−3.00 to −3.99	Severe drought
≤ -4.00	Extreme drought

[a] Fieldhouse and Palmer (389).

particular location." Palmer notes that arid regions can always use more rain; however, the local agricultural economy has become adapted to prevailing conditions. In a "normal" precipitation year then, the PI is zero in both New Mexico and New York State. The word "drought" should always be used in a relative sense because accepted terms such as "desert" and "arid region" describe locations with continual water shortages. Palmer particularly dislikes the phrase, "permanent drought." The following terms have been defined:

(a) *Potential recharge* PR: The amount of moisture required to bring the soil to field capacity.

(b) *Potential loss* PL: The amount of moisture removed from the soil if rainfall were zero during the averaging period.

(c) *Potential runoff* PRO: The amount of runoff that would occur if PET were zero and rainfall equaled the available soil moisture capacity W.

(d) *Climatically appropriate for existing conditions* estimate CAFEC: This is a normalization technique that yields an adjusted estimate of ET, for example, by the following defining equation:

$$(11.17) \qquad CAFEC(ET) = (\overline{ET}/\overline{PET}) \cdot PET$$

Similarly, there are CAFEC values of recharge R, loss L, runoff RO, and precipitation P. In Kansas during June, for example,

$$\overline{ET} = 9.37 \quad cm \quad (3.69 \quad in.), \qquad \overline{PET} = 13.20 \quad cm \quad (5.20 \quad in.)$$

If a particular June had an above-average PET, say 15.24 cm (6.00 in.), then

$$CAFEC(ET) = (9.37/13.20) \, 15.24 = 10.82 \quad cm$$

(e) *Moisture anomaly index* Z:

$$(11.18) \quad Z = K[P - \{CAFEC(ET) + CAFEC(R) \\ + CAFEC(RO) - CAFEC(L)\}]$$

where K is a coefficient which is determined empirically, and which varies with location and season.

Finally, the Palmer index is obtained by examining very dry spells at a number of locations.

$$(11.19) \qquad PI_i = k_1 PI_{i-1} + k_2 Z_i$$

where the coefficients k_1, k_2 vary with averaging time but not with location or season. The subscripts i, $i - 1$ refer to consecutive intervals of time.

The precise details are given in a monograph by Palmer (388). The entire superstructure is based on Thornthwaite's method of estimating PET, and therefore may be criticized. However, the class limits for the index seem to agree with actual meteorological drought reports in the United States. An illustrative calculation averaged by months over parts of seven states between 1961 and 1966 is shown in Fig. 84 (386).

Palmer (390) recognizes that a meteorological drought is not always the same as an agricultural one. The latter is a transpiration deficit (normalized for location and season) whereas the Palmer drought index can be negative because of negative abnormalities in the terms of the hydrologic water budget. The United States "Weekly Weather and Crop Bulletin" therefore includes values of a weekly *crop moisture index* CMI, which is a measure of cumulative transpiration deficit. Palmer (390) has described the way in which CMI is calculated.

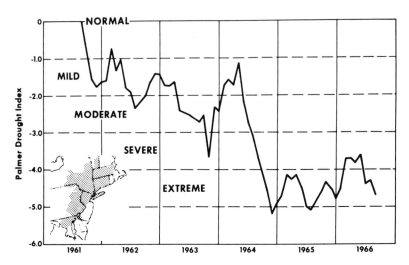

Fig. 84. Drought severity (areal composite drought index) as given by the Palmer drought index for northeastern United States in the years 1961–1966 (386).

Droughts are of interest to statisticians and to dynamic meteorologists. Yevjevich (39) believes that dry periods should be studied from the standpoint of theories of extreme values and of runs. First, however, rational methods must be developed for deciding when a drought has begun or ended.

Anomalies in the general circulation provide clues regarding the causes of droughts. Namias (391), for example, has found that the 1962–1965 drought in the northeastern United States was associated with below-normal temperatures (implying below-normal ET) and with an anomalous northwest component of the upper winds along the Atlantic seaboard. The dynamic climatologist cannot yet predict prolonged droughts but he is beginning to understand the associated links and "feedback" mechanisms.

11.5. AGRICULTURAL YIELD PREDICTION

Crop yield is dependent partly on seasonal weather. The relationships are difficult to unravel, however, because other factors are important. Furthermore, the meteorological variables and the relevant averaging times vary from year to year; sometimes the occurrence of a single rare event such as a hail storm is the limiting stress; in other years, the soil moisture three months prior to harvesting is critical.

The investigator must recognize the existence of weather-sensitive factors, which produce significant correlations between yield and particular meteorological elements but which do not participate in a direct cause–effect relation. Grasshopper infestations are most likely in hot weather while certain types of rusts and molds multiply during cool damp conditions.

Furthermore, there is the human element, commonly referred to as *technology*. Because of improving agricultural practices, long-term data on productivity invariably show an upward trend. Human intervention includes irrigation, the development of drought- and rust-resistant strains, the introduction of modern cultivation practices and shelter belts, and the application of fertilizers and insecticides.

Multivariate analysis is used frequently to seek relationships between productivity and meteorological precursors. Although the results may be statistically significant, they are rarely of more than local value. Furthermore, the regression coefficients often change and must be re-computed periodically as more data become available. Despite these shortcomings, the analysis may lead to physical insight into the relative importance of various meteorological factors and may assist in the selection of the most appropriate averaging and lag times. Several examples illustrate the method.

Sen *et al.* (392) have studied the influence of climatic factors on the yield of tea in the Assam valley in India. Mean values of rainfall, relative humidity, sunshine, diurnal temperature range, and temperature were tried as predictors. A separate analysis was undertaken for each of the early, main, and late crops (April–June, July–September, October–December). Time was added as a predictor to account for changes in growth rate of the tea plant as it ages. After some initial trials, the logarithm of rainfall was used in place of rainfall itself, an increase in precipitation proving to be more beneficial when rainfall was low than when it was high. The investigators found that for the early crop, the significant predictors were mean temperature and the logarithm of rainfall, each with a lag time of three months; the yield was greater after warm, wet weather (up to about 18 cm of rain) than after cool, dry conditions. For the main crop but not the late one, the temperature and logarithm of rainfall in the January–March period were significant predictors. These results provide a basis for physiological studies.

Changnon and Neill (393) have examined the joint effect of weather and technology on corn yields in central Illinois during the years 1955–1963. The study was based on agronometric observations from 60 farms located in a flat area of 1000 km² (400 miles²); meteorological data for each farm were estimated by interpolation between values from a mesoscale network of 49 weather-observing stations. Table XXXVI

TABLE XXXVI

LINEAR CORRELATION COEFFICIENTS[a]

	Corn yield		
Variables	Individual observations	Yearly mean values	Year (trend)
Year	0.70	0.89	—
August mean temperature	−0.69	−0.89	−0.69
Sum of degrees above 90°F in July–August	−0.51	−0.66	−0.61
July mean temperature	−0.50	−0.64	−0.58
Nitrogen	0.41	0.91	0.87
Number of days above 90°F, June–August	−0.37	−0.47	−0.31
Plant population	0.35	0.87	0.97
August rainfall	0.29	0.37	0.30
June mean temperature	0.29	0.37	0.33
July rainfall	0.24	0.34	0.15
June rainfall	−0.22	−0.29	−0.35
Preseason precipitation	0.13	0.22	0.14
May mean temperature	−0.07	−0.09	−0.13
Soil productivity	0.05	—	—
Planting period	−0.01	−0.25	−0.42

[a] Changnon and Neill (393).

summarizes some of the results. The first vertical column tabulates the correlation coefficients between yield and various predictors when observations from the 60 farms and all 9 years were used. In the second column, the correlation coefficients are based on annual mean values averaged over all 60 farms. The third column gives the correlation between each predictor and each year, and thus is a measure of trend. The top horizontal row shows the correlation coefficient between corn yield and year.

There is an upward trend in corn yields and the technology predictors: nitrogen application and plant population. Some of the meteorological variables, on the other hand, show a downward trend through the particular 9-year period. The best predictors of yield were the two technology factors; among the meteorological variables, the best indicator was August mean temperatures; rainfall did not seem to be a limiting factor, perhaps because the 1955–1963 period in Illinois was particularly favorable, with no prolonged droughts. There may indeed have been too much precipitation in some cases.

Huff and Changnon (394) have used the regression equations developed for the Illinois network to simulate the effect on corn yields of increasing the rainfall by cloud seeding over a 14-year period. In three of the years, additional rain would have increased yield; in another four years, there would have been a gain for small increases in rainfall but a reduction in yield if rainfall had increased by more than 60%; in the remaining seven years, cloud seeding would always have reduced yields. The overall result of indiscriminate yearly seeding would have been a small net decrease in corn production.

Long-range methods of forecasting winter wheat yields in the black-soil regions of the Soviet Union have been developed by Ulanova (395). Only two predictors are used: the number of sprouts per square meter and the available soil moisture S in millimeters in the 0–100-cm layer, both observations being made in spring. The yield Y is in centers per hectare (1 center = 100 kg).

(a) For 1000 to 2000 sprouts per square meter:

(11.20) $Y = 0.24S - 10.2, \quad r = 0.86$

(b) For 400 to 900 sprouts per square meter:

(11.21) $Y = 0.2S - 11.1, \quad r = 0.89$

Lowry (396) has studied the meteorological influences on the annual cone production by Douglas fir over a 48-year period in western Washington and western Oregon. The observations he used were given on a qualitative scale: failure, light, medium, and abundant, to which Lowry assigned the numbers $-3, -1, 1, 3$. The data-collection techniques had been reasonably objective and had not changed over the years.

Lowry chose the climatological station at Salem, Oregon as representative of the area. Monthly mean temperatures and precipitation amounts for the 48-year period were divided into quartiles within each of the 12 calendar months and were ranked according to the same scale as yield. Cross products S_T were then formed between the annual yield Y and the monthly mean temperature T, separately for each of 30 monthly lags, where October was month zero (the time of year when the cones matured). Similar cross-products S_P were obtained with monthly precipitation P, i.e.,

(11.22) $$S_T = \sum_i Y_i T_j \quad \text{and} \quad S_P = \sum_i Y_i P_j$$

where

$j = 0, 1, 2, \ldots, 30$ months $i = 1, 2, 3, \ldots, 48$ years

Knowing the joint sampling distribution of S, it is possible to assume the null hypothesis and calculate the values for 5 and 1% significance. The

results are given in Fig. 85. A large positive value of S suggests that a
high yield of cones is associated with above-average temperature or
precipitation. There is a suggestion that the Douglas fir "remembers"
the weather for as long as 27 months prior to ripening of the cones.

A result that is significant at the 5% level may arise by chance once
every 20 times on the average. There is always a risk, therefore, in
accepting the reality of one or two peaks among a relatively large group
of otherwise nonsignificant results. To explore this possibility, Lowry
divided the record of Y into two groups of 24 years each. He did not use
year 24 as the dividing point because more recent observations may
have included more data from high-elevation sites than did the earlier
ones. He also decided not to place alternate years in the two subsamples
in case a 2-year physiological periodicity existed. Instead, each adjacent
pair of years was randomly assigned.

Significant results are always more difficult to attain when a sample
size is halved. Nevertheless, each subclass yielded a significant association with high precipitation at about month 18 or 19 (the spring of the
previous year). For temperature, one subsample showed a positive association at month 9 (January) while the other revealed a negative association at month 27.

Lowry emphasizes that although there is strong evidence that a high
yield is associated with above-average precipitation in the spring preceding the year of harvest, there need not be a direct cause–effect relation.
The governing meteorological variable may be cloudiness, humidity,
wind, or some combination.

These results suggest but do not prove a connection between yield
and year-to-year variability in climate. They do, however, assist in
planning more detailed experiments.

Arakawa (397) has analyzed a long record (1883–1954) of annual
rice production in the Tohoku district of Japan. Figure 86 displays the
yields and the July–August temperatures averaged over six representative stations. Despite random variations, there is an upward trend in
yield due to improving technology. An inspection of harvest and temperature data shows also that crop failures occurred only in cold
summers, notably 1884, 1902, 1905, 1913, 1931, 1934, 1941, and 1945.
Arakawa notes that rice is tropical in origin and requires a summer
temperature of not less than 20°C. Tohoku district is in the northern
part of Japan (37–41° N), where temperature is a sensitive indicator of
yield; farther south, this meteorological element might never be a
limiting stress.

An example of small-scale experimental design has been given by
Pelton (398), who investigated the effect of a windbreak on wheat yields
over open prairie in Saskatchewan, Canada during the years 1960–1964.

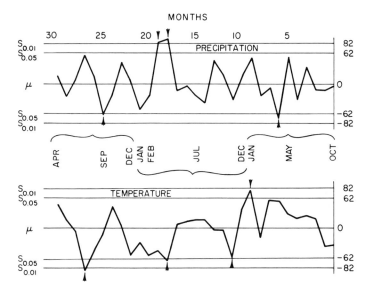

FIG. 85. Values of S for lags of from 0 to 30 months, for precipitation (upper) and temperature (lower). Arrows indicate values beyond the 5% significance level (396).

Because winter snowfall accumulations around obstacles create spatial differences in soil moisture during the subsequent growing season, and because the object of the study was to isolate the effect of wind, snow-fencing was erected after the snow had melted in the spring of each

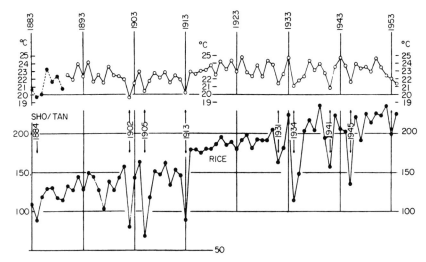

FIG. 86. Relation between the yield of rice in the Tohoku district, Japan, and July–August temperatures 1883–1954. The rice yield is in sho/tan (1 sho/tan = 0.2024 bu/acre) (397).

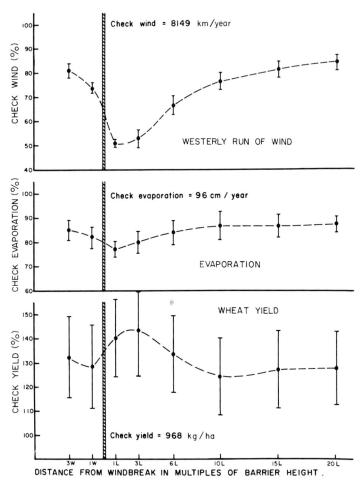

Fig. 87. Effect of a windbreak on run-of-the wind, atmometer estimates of evaporation, and wheat yield, over open prairie near Swift Current, Saskatchewan (398).

year; the dimensions of the fence were 76.2 by 2.4 m, with a porosity of 45%; the orientation was at right angles to the prevailing wind (westerly 56% of the time). The results given in Fig. 87 include wind at a height of 1.2 m, empirical evaporation estimates obtained from atmometers, and wheat yield. Each variable is expressed as a percentage of its value at an undisturbed check point 100 m northwest of the fence. The three curves require little comment; a reduction in wind speed is associated with increased yield. Quite naturally, the wind was not always westerly; thus, the meteorological elements and the yield were modified on both sides of the fence.

Soil moisture is one of the most significant determinants of agricultural yield in many parts of the world. A number of investigators have included a simple index of soil moisture but Baier and Robertson (399, 400) have demonstrated the usefulness of their more comprehensive method in which the moisture is partitioned into 6 layers and according to different stages of crop growth (see Section 11.3). The moisture zones and crop-development periods defined in Table XXXVII provided 24 variables for a multivariate analysis of the yield of Marquis wheat.

Using a simple correlation method, the soil moisture content in the first period P–J (planting to jointing) was not correlated with yield (variables 1–4 in Table XXXVII). In the second and third stages (J–H and H–M), all correlation coefficients were significant at the 1% level. Later in the season, however, the effect of soil moisture on yield decreased in the upper zones of the soil but increased at root depth.

Wheat yields were defined in three ways—number of heads per row, number of kernels per head, and 1000-kernel weight in grams. Figure 88 shows the relation between observed yields and the predictions obtained by multivariate analysis of the five most important soil moisture variables. In view of the wide diversity of climate and photoperiod, the agreement is excellent; Fort Simpson is at 62° N latitude while Harrow is at 42° N.

A number of investigators have considered biological responses, in which weather plays an indirect but nevertheless significant part. Two examples are honey production and the forecasting of milk yields.

Smith (401) examined the weekly milk totals in England and Wales for the period 1953–1966; the data represent the mean performance of almost a million dairy cows. The yield displays an annual cycle, reaching a maximum in late spring and a minimum in late autumn. Not unexpectedly also, there is an upward trend over the 14 years, reflecting the influence of technology. When this factor is included in the regression equation, Smith finds that a 12-month (April–March) forecast of yield can be made with some skill at the end of March, based on the production data for the previous November–March period and on the March soil temperature at Oxford (a representative country location). The forecast can be improved at the end of June by using milk yields for April–June and the rainfall in June; this latter quantity has an important effect on the quality of the summer hay, the principal food in the following winter.

Hurst (402) has related annual honey production in Britain to summer temperatures. The data are for the years 1928–1966, and no technology trend exists; in fact, the yield per colony of bees has declined slightly, which parallels a decrease in mean summer temperature over the same period. Hurst finds a correlation of 0.59 between yearly honey production and the August mean temperature anomaly. Assuming random

TABLE XXXVII

DEFINITION OF SOIL-MOISTURE ZONES, CROP-DEVELOPING PERIODS AND LISTING
OF SOIL-MOISTURE VARIABLES (BODY OF TABLE)[a]

Zone	Percentage of total capacity	Crop-developing period[b]				
		P–J	J–H	H–M	M–S	S–R
1	5.0	1	5[c]	9[c]	13	19
2	7.5	2	6[c]	10[c]	14[d]	20
3	12.5	3	7[c]	11[c]	15[c]	21
4	25.0	4	8[c]	12[c]	16[c]	22[c]
5	25.0	—[e]	—[e]	—[e]	17[c]	23[c]
6	25.0	—	—	—	18[d]	24[c]

[a] Baier and Robinson (399).
[b] P = planting; J = jointing; H = heading; M = milk stage; S = soft dough;
R = ripe.
[c] Significant simple correlation at the 1% level with yield.
[d] Significant simple correlation at the 5% level with yield.
[e] Below root depth.

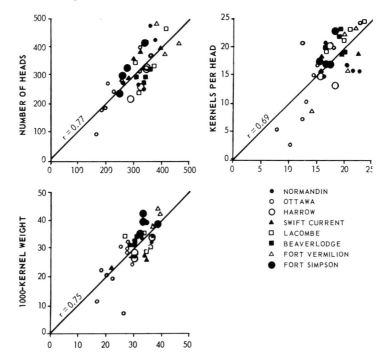

FIG. 88. Relationship between crop yield components observed (ordinate) and
derived (abscissa) from estimated soil moisture for eight stations across Canada
during the years 1953–1957 (400).

sampling theory (not justified), the coefficient is significant at the 0.1% level.

11.6. SEASONAL STUDIES OF HEALTH

Accident rates are weather dependent and thus seasonally dependent: drownings are most likely in summer whereas traffic accidents are most frequent during fog and freezing rain. Many common diseases also show seasonal cycles but there is no obvious explanation in most cases.

As a particular example, a number of investigators have suggested that ulcers are aggravated by temperature changes. This view is supported by the fact that there is an increased incidence of reported cases of ulcers in Philadelphia during the months October–March and in southern Australia during May–July. A meteorological connection seems obvious but the hypothesis has been challenged by Tom et al. (403) who examined the seasonal trend in Hawaii. In that part of the world, the annual temperature range is small and there is no cold-weather stress. However, these investigators found a seasonal cycle in ulcer incidence, rather similar to that in Philadelphia, but bearing no relation to either monthly or interdiurnal temperature change. This negative result is valuable, and the methodology is to be noted. Without such evidence it would be perfectly natural to recommend that a "warm climate with relatively little daily and seasonal change in temperature would afford the most suitable residence for those prone to suffer from duodenal ulcers" (403).

There is a vast literature on medical meteorology. In many studies the evidence is inductive and not very convincing. There are, of course, exceptions in which physiological links have been established. The annual cycle of hay fever can be explained readily, as can the local incidence of skin cancer, which is related to latitudinal and altitudinal variations in the intensity of solar radiation. In all cases, a recommended approach is cross-spectral analysis of daily medical data with daily averages of air quality and of meteorological elements or indices; this method has been used rarely, if at all.

The volumes by Landsberg (258) and Licht (404) are recommended for further reading.

11.7. POLLUTION STUDIES

The simplest kinds of pollution measurements are obtained with deposit cans for dustfall and with lead peroxide candles for SO_2. These yield monthly values, a time period during which the weather varies greatly. The data are quite satisfactory for defining seasonal and long-term trends in air quality, but they do not indicate whether the trends are due to changes in source strength or in the atmospheric general circulation. Why, for example, were dustfall totals so different in two consecutive Januarys?

Another application on this time scale concerns design studies for proposed power plants or large factories. The safety analysis requires climatological predictions of the concentrations of pollution. Because of the averaging problem mentioned in connection with Eq. (2.4), little is to be gained from a knowledge of mean monthly wind speed and direction. Instead, an analysis of hourly data is recommended, in which the wind is specified to one of eight directions. The assumptions are made that the plume during each hour is contained within one of the eight 45° arcs, and that the center-line concentrations at various distances from the source may be estimated by standard methods (16, p. 209, for example). Although part of a plume may overlap into another arc, particularly when the wind direction is changing, gains and losses to adjacent arcs are assumed to balance, approximately. The mean monthly concentration at any point is then given by summation.

If daily Grosswetterlagen are readily available, the method of Schmidt and Velds (405) is an attractive alternative. These investigators used Velds' results (described in Section 10.2) to explain the improving winter-time air quality in Rotterdam during the period 1962–1968. Some results given in Table XXXVIII show that the frequencies of HM, HN_a, and Sz types have decreased. These types are associated with above-average pollution levels. The reduction in SO_2 concentrations can there-

TABLE XXXVIII
Comparison of 6 Winters in Rotterdam[a]

Winter	Number of days with				Mean precipitation (mm)	Mean wind speed (m/sec)	Mean temperature (°C)
	HM	HN_a	Sz	Total			
1962–63	32	10	13	55	304	5.0	2.2
1963–64	35	13	3	51	269	5.1	4.5
1964–65	13	7	2	22	472	5.6	4.9
1965–66	26	3	0	29	492	6.5	4.9
1966–67	9	0	2	11	498	6.7	6.6
1967–68	8	0	0	8	468	6.8	5.8

[a] Schmidt and Velds (405).

fore be explained meteorologically and there is no reason to be optimistic about future long-term trends in air quality.

In conclusion, a time series of normalized annual pollution concentrations may be derived, in which the effect of year-to-year weather variability is removed. The first step is to establish relations between air quality and meteorological factors, using pollution wind roses as in Fig. 45, or urban multiple-source computer models (323). Next, one year is chosen as reference. Finally, the weather frequencies in other years are adjusted to those of the reference year, permitting calculation of normalized annual pollution concentrations.

12/STUDIES OF PAST CLIMATES

12.1. INTRODUCTION

Homogeneous geophysical time series contain red noise. This fact has intrigued climatologists for many years and has led to countless studies of trends and anomalies. There are three objectives in such investigations:

(a) To locate and "clean up" very long time series.
(b) To identify nonstationary features.
(c) To seek physical explanations for trends and anomalies.

The existence of "unusual" historical periods seems to be well documented; for example, there was a warm spell in northern Europe from about 1000 to 1300 AD and a "Little Ice Age" between 1550 and 1700 AD. Further back in time, a thermal maximum occurred about 3000 BC. In this connection, the notation BP (before the present) is sometimes used instead of AD and BC; thus, 3000 BC is equivalent to 5000 BP.

A number of physiological and ecological problems require many years of observations before there is hope of obtaining significant results. For instance, an urgent need exists for information on long-term trends in global air quality. Clearly a sampling program should begin at once and continue for many decades, but extrapolation back in time would also be useful in order to estimate the effect of the industrial revolution and of the automobile era on world-wide pollution levels.

In this and other applications, historical data are fragmentary and often of doubtful quality. Yet in a study of the ecological effects of the "Little Ice Age" in the seventeenth century, the experiment cannot be repeated. There is therefore a continuing search for an improved climatic calendar using ancient writings, rocks, glaciers, bogs, oceans, and every other conceivable tool. Sections 12.2–12.4 describe methods which have been used or which appear to be promising. The point to be emphasized is that knowledge of past climates is not approaching a limiting state by any means. Many indicators are locked up in the earth's crust, awaiting improvement in measurement techniques. Other data already exist but are diffused through various scientific disciplines. Sheppard (406) has recommended that the various types of evidence be arranged in uniform quantitative form. Only then can meaningful problems be formulated and solved.

12.2. THE INSTRUMENT ERA

The barometer was invented in 1643. Crude rainfall gauges were tested at about the same time while daily readings of temperature began in England in 1664. The seventeenth century can therefore be said to usher in the instrument era of meteorology. As might be expected, however, early data are difficult to interpret. Manley (407) has produced a painstaking estimate of monthly mean temperatures in central England as far back as 1698. Some of the problems he faced were the changeover in 1752 from the Julian to the Gregorian calendar, the exposure of most thermometers in unheated rooms instead of outdoors until 1755, and the use of an early thermometer scale on which the freezing point was between 78 and 79 while 88°F was represented by zero. Also, in the case of barometric pressure (17, p. 8), some eighteenth century records have been found to contain monthly mean values that are the arithmetic average of the highest and lowest readings in the month.

Mitchell et al. (17, p. 32) define the secular period as that historical period for which there are "quantitative meteorological observations at a number of geographically well-distributed observing stations." These authors state that "in most parts of the world the secular period is considered to have begun not earlier than the early or mid-nineteenth century." The rapid increase over the last century in the number of observing stations in the United States and possessions is shown in Fig. 89 (408).

For many investigations, the only meteorological data of practical value are those on punched cards or magnetic tape, which are readily retrievable from National Meteorological Services. In Canada, for example, hourly data are "archived" in this way only as far back as 1947. Daily values of maximum and minimum temperature and of rainfall, snowfall, and total precipitation are, of course, available on punched cards for longer periods (back as far as 1845 for Toronto).

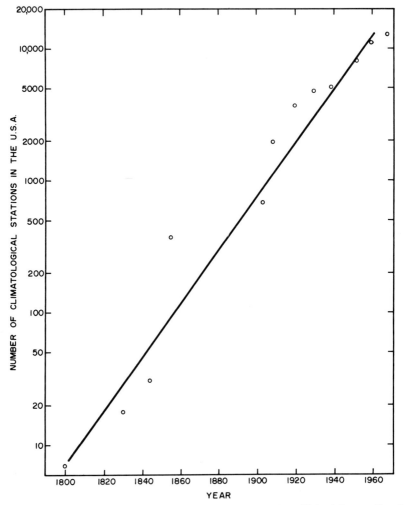

FIG. 89. Number of climatological stations listed in various United States climatic surveys since 1800. The ordinate is on a logarithmic scale (408).

Even for these relatively short lengths of record, care must be taken in using the data. Changes in synoptic codes continually tax the ingenuity of computer programmers. International agreement was reached in 1965 to report wind direction to 36 rather than to 16 points. Other revisions such as those relating to cloud opacities (in the early 1950's) are pitfalls awaiting the investigator who uses a deck of punched cards.

The regular climatological data may be supplemented in space and time with various meteorological and biological indices. Some of these will be discussed in Sections 12.3 and 12.4. It is to be noted here, however, that the indices must be "calibrated" against regular weather observations if they are to be used with any confidence at all. This is a central problem in biometeorology. Indices such as the dates of freezing of a river or the diameters of tree rings integrate the effects of environmental variables over long intervals of time. In order to unravel interactions, carefully designed experiments are required over short time periods. Only then can historical indices be used with reasonable assurance.

12.3. THE HISTORICAL PERIOD

Weather information for the historical period includes data from various written sources, as well as regular climatic data. Historical documents often include explicit or implicit references to weather. As a primary source, there are a number of excellent weather diaries, as well as official summaries of the general character of summers and winters, dating back in Europe to about 1100 AD (409). [An example of an eighteenth century weather code is given in Fig. 90 (410).] As a secondary source, there are long records of phenological events such as the yearly freezing date of Lake Suwa in Japan since the winter of 1443–1444 (103), the incidence of ice along the coast of Iceland back to 940 AD (411, p. 80), and the times of harvest (in various church and other documents in a number of countries). Third, inferences can be made from census information. For example, the *Domesday Book* of 1085 indicates that there were at least 38 vineyards in England at that time, suggesting a warmer climate than exists today (212). Thomas (412) has used the *Acreage Returns* of 1801 to relate crop distributions in Wales to the mean annual rainfall. Finally, cultural and economic evidence may be used as climatic indices provided that independent methods yield similar results. Examples include migrations of prehistoric tribes, trends in prices, population statistics, and annual exports of agricultural products. In these cases, the time constant of the response must be considered. In tension areas such as Greenland and Arizona, ecological reaction to weather anomalies is more rapid than in central Europe.

⊙ SUNSHINE   ℮℮℮℮ WIND (PRESUMABLY STRONG WIND)

ᴧᴧᴧ CLOUD   ⠿   RAIN

FIG. 90. The weather symbols used by Dr. Claver Morris, a Somerset physician, who kept a weather diary between 1684 and 1726 (410).

Sifting and weighing of such data is not without difficulty. Oliver (413) suggests that weather diaries may emphasize rare events unduly and may be biased if a tenant is seeking a reduction in his rents because of poor crops. Explorers' notes are not very helpful, time and space variations being intertwined.

Independent overlapping data are essential, as a cross check on the tentative climatic calendar and as an indication that the events are of regional as well as local significance. In hilly country, mesoscale influences may be predominant. Oliver (414), for example, notes that the mean monthly temperature anomaly for a valley station may have the opposite sign to that on the uplands at times when the mean pressure pattern is abnormal.

At a conference on the climate of the eleventh and sixteenth centuries (411), the importance of independent data verification was emphasized. Summer wetness values presented by Lamb (411, p. 88) for England for the decades between 1210 and 1349 were compared with those derived by Titow using audited accounts from manors owned by the bishopric of Winchester. The two sets of independent data yielded a correlation coefficient of 0.77, significant at the 95% level. Winter anomalies in precipitation presented by Lamb and separately by Herlihy, on the other hand, revealed discrepancies (411, p. 38). As noted in the conference summary, "Whatever the cause of the discrepancy, the very fact of disagreement points out the need for historians to discriminate carefully among the various kinds of meteorological phenomena, and for meteorologists to understand some of the complexities involved in working with original documents."

When regional anomalies are suggested by independent data, they may be used to develop physical models (see Section 12.5). In the tenth century, for example, northern Europe was warmer than usual but there is evidence of cold weather in the Mediterranean region, with ice on the Tiber and the Nile rivers on more than one occasion (212). This leads to a plausible hypothesis about the mean pressure patterns during that period.

There is an interdisciplinary requirement for a climatic calendar. Archaeologists cannot interpret their findings without a knowledge of environmental conditions prevailing in distant times. Historians, anthropologists, sociologists, historical geographers, and historical economists all must consider whether climate or climatic change is significant in the interpretation of past events. Even today, climatic trends "must remain one of our greatest anxieties" (415), particularly where conditions are marginal as at the edges of deserts or at tree lines. Manley (415) suggests that a prolonged period of strengthening of the Scandinavian and Hudson Bay high pressure areas would have a far greater effect on British economy than would any change in tariffs.

Historical climatology has hardly tapped the store of documents from earlier civilizations. However, more interdisciplinary conferences are desirable to exchange results and to seek uniformity in methodology. Furthermore, an international repository for data is recommended (411, p. 82). The analysis of an isolated document from early times may not be worth publication. However, it may prove to be a vital link in a chain of circumstantial evidence concerning world climate during a particular historical period.

12.4. PALEOCLIMATOLOGY

Paleoclimatology is the study of past climates on all time scales, using every possible indicator. The primary experimental methods may be considered under the following general headings:

(a) Climatology of the instrument era (Section 12.2).
(b) Historical climatology (Section 12.3).
(c) Radiocarbon dating.
(d) Tree-ring analysis (dendroclimatology).
(e) Pollen studies (palynology).
(f) Glaciology and snow-line determinations.
(g) Studies of oceans and lakes.
(h) Global vegetation distribution in ancient times.
(i) Paleogeology (sedimentation, magnetism, fossils, etc.).

No method is clearly superior. Instead, the various pieces of geophysical and biological evidence must be compared in an attempt to infer coherent pictures of past climates. Methods (c)–(i) will be described briefly in this section.

Radiocarbon Dating

Atmospheric carbon dioxide contains the isotope ^{14}C, produced by secondary collisions in cosmic-ray showers. Plants and animals convert this CO_2 into organic material (Section 8.5), but the process ceases when the organism dies. Furthermore, so few cosmic-ray particles reach the ground that atmospheric CO_2 is the only significant source of ^{14}C in the world.

This isotope is estimated to have a half-life of 5730 ± 40 years, although there may still be some uncertainty (416). This means that with certain assumptions, radiocarbon dating may be used to determine ages up to about 50,000 years. The method is undoubtedly reliable for organic material formed in the last two centuries. Earlier dating requires the assumption that the ^{14}C atmospheric concentration has remained constant in time and space, i.e., that there has been no change in the balance of primary production of ^{14}C and air/sea exchange rates of CO_2. In fact, by using material of known age as an absolute standard (417), there is increasing evidence that atmospheric ^{14}C concentrations have fluctuated by 2–3% over historical times, resulting in an error of about 1000 years in the determination of the age of an 8000-year object, for example.

Lamb (418) suggests that the difficulty be exploited to produce a climatic index, i.e., the fluctuations of ^{14}C atmospheric concentrations may be inferred from the corrections to the radiocarbon estimate required to obtain the ages of known "standards." A standard might be, for example, a timber from a house known to be built in a particular year. If the index is used as a basis for physical models, there remains, of course, the problem of deciding whether a change in ^{14}C is due to a variation in solar activity or to an anomaly in the exchange rate of CO_2 with the oceans, through the mechanism of a change in either sea temperature or rate of overturning of the oceans.

Radiocarbon dating is used also to determine the age of archeological remains, which in turn may shed light on historical climatology.

Tree-Ring Analysis

Tree-ring diameters are highly correlated with winter precipitation near the edges of deserts in the temperate zones, and with summer temperatures along the arctic timberline. Elsewhere, environmental relationships are much more complex, particularly in the tropics.

The determination of a reliable time-series index requires careful experimental and interpretive work. Rings may be missing or on only one side of the stem while there may be "false" annual rings in other cases. In addition, it is essential that a number of trees show the same ring-width patterns; otherwise, climate is probably not the limiting growth factor.

The height of greatest trunk growth is just below the active crown. Most cores are cut at breast height, above the region of maximum growth in a young sapling but well below it in later years. In addition, the growth rate just below the crown (and at other levels too) decreases with time because of ecological competition and other factors. Thus, a normalization technique is desirable. The first step is to obtain a smoothed growth curve, either by a least-squares fitting of an exponential function or by averaging ring widths for many trees (419). In the latter case, the averaging is by tree age rather than by calendar year, and the sample includes trees of many different ages to eliminate meteorological effects.

Let W_i be the smoothed growth in the ith year, where i is tree age, and let X_i be the actual growth. Then the quantity G_i is defined as follows:

$$(12.1) \qquad\qquad G_i = (X_i - W_i)/s$$

where s is the standard deviation of all $(X_i - W_i)$. Clearly G_i has a mean value of zero and a standard deviation of unity. The quantity is an index of the weather in year i.

Examination of tree-ring series reveals that at first, $W_i < W_{i+1}$ but later, $W_i > W_{i+1}$. An example in Fig. 91 shows that for black spruce in northern Manitoba, the tree-ring width increases annually only for the first 10 years (419). The later portion of the smoothed growth-rate series, for which $W_i > W_{i+1}$, is called the *inverse range*. Only this part of the series is used.

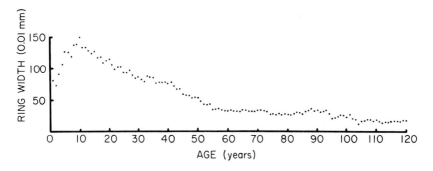

FIG. 91. Growth curve for tree-ring width for black spruce for The Pas, Manitoba, Canada (419).

Mitchell (419) notes that there is a tendency for the standard deviation s in Eq. (12.1) to decrease with age. The resulting nonstationarity in G_i may be removed empirically by using a separate value of s in each consecutive 20-year period. The correction, however, may make the series unsuitable for study of climatic trends.

An informative survey paper on the use of tree rings in the study of climate (*dendrochronology*) has been written by Fritts (420). Some tree-ring chronologies date back more than 7000 years.

Pollen Studies

Palynology is the study of pollen, spores, and related microscopic parts of plants, including fossil material. *Bog stratigraphy* is the analysis of cores to obtain vertical profiles of pollen and spores laid down through the centuries. Radiocarbon dating provides an estimate of the ages of different layers while a count of the species in each layer yields information about the ecology of the period.

Some plants are more sensitive than others to climatic stress. Unfortunately, the best indicators are not always those that produce enough pollen to leave a record. One plant that meets both criteria and has been studied widely in Europe is ivy. Its distribution in historical times, however, is affected not only by the weather but also by human activities. An anthropological explanation has also been given for the decline of tree pollen after about 3000 BC in Europe when man began to clear the forests and develop a shepherd–farmer culture (421).

Pollen studies cannot be used for determining the fine-scale structure of climatic fluctuations. The response of pollen populations to local change is damped by distant transport and by the existence of anomalous meso- and microclimates which remain favorable for growth despite regionally unsuitable weather. Nevertheless, the method is a useful indicator of large-scale effects over time intervals of about 1000 years extending back to about 50,000 years.

Glaciology and Snow-Line Determinations

Glaciers represent a continuously changing net difference between precipitation and evaporation/runoff over many centuries. Numerous studies have attempted to relate the retreat and advance of glaciers to meteorological factors but the observations are difficult to interpret and are not yet a particularly useful climatic index. Manley (422) has emphasized that prolonged anomalies in the atmospheric general circulation may cause some glaciers to grow and others to weaken. Examination of the "synoptic" behavior of a large number of glaciers around the world might prove to be useful. However, the way in which ice slides along its bed is not fully understood, nor is there adequate documentation of the time lag between a climatic and a glacial change; the shape of the basin or plateau seems to be important.

Snow-line determinations can be made but again there is an interpretative problem, with no easy way of disentangling the relative influences of winter precipitation, summer temperature, and physical motion of the glacier. A tool that has been used in these studies is *lichenometry*. The growth rate of certain types of lichens can be estimated (about 0.40

mm/year), permitting inferences to be made about past positions of the edge of the ice sheet, when the glacier is retreating.

Another possibility is examination of vertical profiles in the glacier. Physical and chemical analyses of cores have shown, for example, an order-of-magnitude increase in lead concentrations in Greenland over the last 30 years. The extension of this work to very deep cores seems promising. Radio echoes have been obtained in Northern Greenland, not only from the solid rock beneath the ice but also from a lesser depth of about 400 m (secondary echoes) (423). Evans (424) speculates that the latter may be due to a change in snow density several meters thick, indicating melting and refreezing over many decades. Extrapolation of present rates of snow accumulation indicates an age of 800–1000 years.

Studies of Oceans and Lakes

Fluctuations in sea level reflect changes in the precipitation/runoff-evaporation balance. At times when more water than usual is locked up in glaciers, the sea level is depressed. However, this is not yet a useful climatic indicator because of the disturbing local influences of tectonic upheavals and of the deposition and compaction of sediment. Fluctuations in lake level are also difficult to interpret because of the changing hydrologic characteristics of the drainage basin.

Sediment at the bottom of oceans contains a record of geological time. Wiseman (425) suggests that the most likely oceanic climatic index is the $CaCO_3$ sedimentation rate, which is related to the plankton concentrations in the surface layers of the sea. The abundance of these tiny organisms depends on water temperature, particularly in regions where other stresses are not limiting. Hendy and Wilson (426) have used the method to determine the paleotemperature over the last 100,000 years for a site near New Zealand. The experimental technique was developed by Emiliani (427).

Sediment at the bottom of lakes is also being studied. Here again, a location in a tension zone near the edge of a desert, for example, is most likely to yield a large response amplitude to climatic fluctuations. Most of the present work concerns pollen and seed analysis but chemical determinations of carbon and halogens are also being undertaken (421). Fluctuations about the mean vertical profiles [using an index rather suggestive of that in Eq. (12.1)] may reflect prolonged anomalies in the precipitation/evaporation ratio.

Global Vegetation Distributions

Whereas pollen are carried by the wind over great distances, the discovery of an ancient tree stump or leaf often assures positive identification of position. This statement must be qualified when the specimen has come from a mountainous region or has been in the path of a moving glacier. Bryson (70) believes that the best indicator of climatic change is the record of retreat and advance of vegetation boundaries on the macroscale. Although data exist on vegetation changes at particular locations, there is a real need for "synoptic" maps representing conditions at historical intervals of 100 or 1000 years.

Paleogeology

For time scales of the order of several hundred million years, palaeogeology is the principal source of inference regarding ancient climates. Runcorn (428) described a method of estimating prevailing "paleowinds," from the orientation of dunes preserved in sandstone. A correction must be applied for geological tilt; in addition, a number of regional determinations must be in agreement to ensure that the direction is of more than mesoscale significance. Finally, the observations must be related to palaeolatitude because the poles and continents have wandered.

For more recent time periods, Starkel (429) believes that certain geomorphological forms such as river terraces and lake beaches are closely related to climatic change. Although precise dating is not possible, the visual observations are useful in supplementing and complementing other data.

12.5. PHYSICAL MODELS

The meteorologist has a compelling desire to infer the general circulation and pressure patterns in distant times from a knowledge of temperature and precipitation anomalies. Scientists in other disciplines may quarrel occasionally that this is an irrelevant exercise (411, p. 76) but it does permit predictions to be made for areas where there are no data. In addition to suggesting regions of the world where the search for climatic indicators is likely to be most fruitful, subsequent verification of the predictions is a test of the internal consistency of the model.

The cause of the anomalous pressure patterns remains to be explained but at least one step in the chain is understood. It is not sufficient, for example, to propose that an increase in atmospheric CO_2 concentrations has caused an upward (or downward) trend in world temperatures. The equations of radiative transfer must be solved, and interactions within the general circulation must be understood.

Lamb (430) has made a substantial contribution to the knowledge of

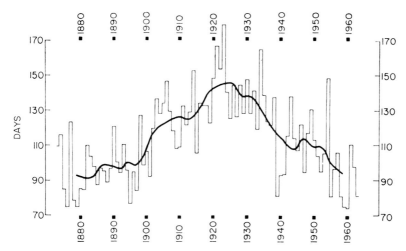

Fig. 92. Frequency of westerly-type days in the British Isles for the years 1873–1963 (430).

pressure patterns during the instrument era by examining early barometer and wind records. His studies indicate that long-term climatic fluctuations (of the order of 50 years or so) are well correlated with changes in the general circulation. Figure **92** shows the frequency of westerly-type days in the British Isles (see Section 10.1) (430). The results help to explain the climatic warming that occurred from 1890 to about 1935. In a rather similar aproach, Hoinkes (431) has suggested that glacier variations in Switzerland between 1891 and 1965 can be related to the frequency of the various classes of Grosswetterlagen (see Section 10.1).

Wahl *et al.* (432) have compared the climate at Portage, Wisconsin in 1829–1843 with that in 1931–1948. The earlier period was 1.9°C (3.4°F) colder than the later period. In order to determine whether this difference was due to local or to large-scale factors, a regression equation was derived between wind direction and temperature for the earlier record. Upon normalization of the wind rose to the 1931–1948 frequencies, the nineteenth century annual mean temperature was brought to within 0.2°C (0.4°F) of the modern mean.

Results such as these, obtained from the instrument era, lend confidence to the hypothesis that past climates are correlated similarly with mean pressure maps of earlier times. This being so, the dynamic climatologist seeks to determine whether these historical "spells" can be explained without invoking the influence of external parameters such as variations in the solar constant, hypotheses which cannot be checked from

historical data. Every year there are periods when the mid-latitude wester-
lies are interrupted by blocking high pressure areas, particularly over
western Europe. An increased frequency of these synoptic patterns
would produce the necessary pressure anomalies but this leads to a
further question: what determines the frequency of blocking anticyclones?

The key to this riddle may exist in data from the modern instrument
era, commencing at the time when radiosondes and upper air maps
became available. Prolonged anomalies of sea temperature, snow and ice
cover, and the latent heat export from tropical continents have been
suggested, for example, as possible triggering mechanisms. But how
large must a glacier be before it begins to affect the general circulation,
if it does so at all? These are difficult questions but few scientists will
disagree with the proposition that meaningful answers can only be found
from a three-dimensional model of the atmosphere.

Curry (53) suggests that major climatic changes may have arisen by
chance. Because of storage terms, particularly in the tropical oceans and
in glaciers, there is no reason to assume that the atmospheric sinks and
sources of energy must balance in any calendar year. Curry speculates
that equilibrium conditions may never be established, and that there is

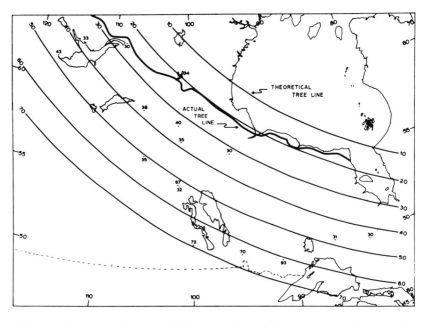

Fig. 93. Mean tree-ring growth lines, actual tree line, and theoretical tree line;
units of mean growth of 0.01 mm (419).

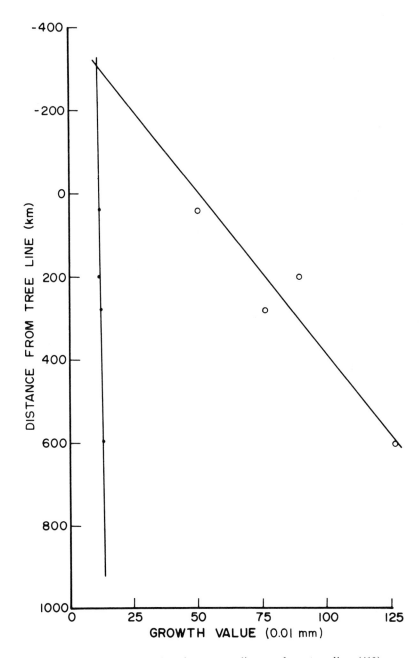

FIG. 94. Tree-ring growth value versus distance from tree line (419).

continual climatic change. He reasons that if the tropical oceans have a 50% chance of acting as a net source or sink in any one year, they can create ice ages from time to time.

Lorenz (433) defines a *transitive* system of equations as a set that results in unique long-term statistics for given initial conditions. He considers the question of whether climate is a transitive or intransitive process, but he concludes that there is not yet sufficient evidence upon which to base an opinion.

As mentioned briefly in Section 3.4, Hare (69) noted that it may be more than coincidence that the arctic tree line and the summer mean surface position of the arctic front lie so close together (see Fig. 23). Bryson (70) has considered the possible links in greater detail; he believes that the key to long-term climatic change may well be found in air-mass analysis. The Department of Meteorology at the University of Wisconsin has a continuing program to test the theory. The research has stimulated physiological and ecological studies of the tree line in northern Canada. Over the tundra near Ennadai Lake in the Northwest Territories, for example, Larsen (434) has found that arctic flora are relatively rare within 240 km (150 miles) of the tree line, although the topography and geology of the region is uniform. There is also some evidence that the forest extended farther north at some distant time.

Mitchell (419) has analyzed tree rings from a number of locations in the Canadian boreal forest, relating growth rates to distance from the tree line. He used the biological ring width of the one hundredth year of the inverse range [see the discussion of Eq. (12.1)], thus eliminating year-to-year meteorological variability. The results given in Fig. 93 show a gradient from northeast to southwest. The "theoretical" tree line was determined in a way illustrated in Fig. 94, and may be explained as follows. Depending on soil types and drainage, there is a considerable scatter in growth rates within the boreal zone. The two regression lines in Fig. 94 represent the 90th percentile of the high-growth sites and the 10th percentile of the low-growth areas. These percentiles were chosen because they are close to the extremes but nevertheless are relatively stable statistics for small populations. The two regression lines intersect at a point corresponding to a distance 320 km (200 miles) beyond the present tree line. Trees should exist in scattered groups in favored locations in this area, as indeed they do.

12.6. SIMULATION OF CLIMATIC CHANGE

As yet, climatic fluctuations cannot be simulated with a model of the general circulation. The *Global Atmospheric Research Program* GARP, planned for the late 1970's by the International Council of Scientific Unions and the World Meteorological Organization, has the more modest goal of testing the predictability of the atmosphere for periods of 2–3 weeks. This in itself is a "quantum jump" forward, requiring larger computers, an increased observational network, and improved modeling of surface boundary conditions. Although a small error in initial or boundary conditions in some remote part of the world has little effect on a 36-hr forecast, the error amplifies over longer times.

The GARP experiment will provide clues about climatic change. For example, external parameters such as snow cover or sea temperature may be varied in a computer simulation; this will test the stability of the general circulation to such changes.

A modest but noteworthy attempt along rather similar lines has been described by Takahashi (435). Using a simplified model of heat transfer, Takahashi examines the effect of a 1% sinusoidal variation in the solar radiation received at the earth's surface. Table XXXIX shows the

TABLE XXXIX

AMPLITUDE OF AIR TEMPERATURE VARIATION (°C) FOR 1% VARIATION OF SOLAR RADIATION RECEIVED AT THE EARTH'S SURFACE[a]

Period (years)	1	10	100	1000	10,000
On sea	0.05	0.14	0.33	0.55	0.70
On land	0.30	0.65	0.73	0.75	0.76

[a] Takahashi (435).

resulting amplitudes of air temperature for different impressed frequencies. The study illustrates the importance of response time. Because of the thermal lag of the oceans, a one-year cycle has hardly any effect on air temperature; only for slow oscillations of the order of thousands of years is there a significant amplitude. The response is more rapid over land, but a large fraction of the world is covered by water. Because of the simplifying assumptions in the model, absolute values of the amplitudes are uncertain. Nevertheless, the relative lags are likely to be representative.

12.7. CLIMATIC DETERMINISM

The theory of climatic determinism holds that climate has had a profound effect on the course of history. Lewthwaite (436) has discussed the broad question of environmental control over man's destiny, a controversial topic for many years. Lewthwaite suggests that the term "environmentalism" has become unpopular in geography because it is used in many different contexts, some of which include metaphysical overtones. Although some historical events were set in motion by environmental factors, there are others that arose by chance.

The example of climatic determinism usually quoted is famine, which undoubtedly has been a factor in the migration of tribes in early times, and in some emigrations more recently. Lamb (212) has speculated that floods in Hungary about 1300 BC brought the Achaeans to Greece. The thickening of the polar ice in the thirteenth century was a calamity for the Viking colony in Greenland, and even today there are many weather-sensitive segments of the economy. A series of crop failures may affect the political stability or the economic policy of a country.

The method of the historian is to examine documents and weigh evidence, drawing upon the resources of the physical and social sciences. After the "facts" have been established, a number of possible explanations are examined in turn. The resulting interpretation of a period in history usually cannot be tested by the scientific method, although the circumstantial evidence may be convincing to others who have studied the data in equal depth.

Even for events in the present decade, it is not always possible to "prove" in the scientific sense that a crop failure has had certain political consequences. For this reason, the theory of climatic determinism has not been accepted very seriously by the meteorologist as a subject for scientific inquiry.

Bryson (437) comments that for every writer who states that the Near East became a desert because of agricultural mismanagement, there is another who takes the opposite view. Bryson believes that the hypothesis is capable of scientific evaluation, and he has described a study in progress in the Rajasthan desert of northwestern India and southeastern West Pakistan, an area of about 10^6 km^2 (a quarter to a half million square miles). Archaeological evidence indicates that in the period 3000–1700 BC, the area was inhabited by the Harappans, who grew grain and raised cows and pigs. There is some indication that in later times, the region was occupied intermittently along river valleys by other cultures, whereas the Harappan sites were on high ground, suggesting that "the floodplains were farmed" (437).

It is a fact that the fertile land is now a desert. What remains to be determined is whether the change is due to natural causes or is man-made. The former possibility implies that a prolonged period of drought initiated irreversible changes. This theory may be examined by study of the paleoclimatology of adjacent areas and by more extensive and more precise archaeological dating in the Rajasthan desert itself; much of the area has not yet been explored.

The principal evidence supporting a man-made change is that this desert is a meteorological anomaly. The precipitable water vapor in the atmosphere is four times as great as that found over other arid regions, yet the rainfall is scanty. The desert is covered by a thick atmospheric blanket of dust, which affects the radiation balance and may contribute to atmospheric subsidence. This possibility is being examined by theoretical and experimental studies of radiative transfer, including numerical modeling.

The investigation is not only of historical but also of practical importance; reclamation of the desert would have far-reaching economic consequences. Bryson and Baerreis (438) are correct in suggesting that integrated studies of the past, the present, and the future are most likely to be fruitful.

12.8. CLIMATIC ADAPTATION

Like environmentalism, the study of climatic adaptation is still confounded by "a mass, or mess, of mythology and speculation" (436). Short-term adaptation to meteorological extremes is now reasonably well understood from controlled laboratory experiments, particularly as related to the effects of heat, cold, and sharply reduced atmospheric pressure. A living organism accommodates as best it can to a harsh environment. Similarly, there have been many laboratory studies of adaptation over several generations. For example, when mice are kept at a temperature of $-3°C$ (439), the nestling mortality is high among the initial population; by the 12th generation, however, the death rate declines to that of mice maintained at 21°C. This occurs both for genetically mixed and for highly inbred strains of mice.

The main area of controversy concerns the question of adaptation to modest stresses that have been acting over decades or centuries. What physiological effects, if any, have taken place during the last 50 years as a result of the introduction of central heating and air conditioning? Because there have been other environmental changes during this interval of time, it is not possible to obtain a definitive answer to the question. Plausible explanations are more likely in situations where one particular stress is dominant. Examples include studies of populations living at high altitudes in the Andes, and a comparison of arctic and temperate zone plants; Larsen (11, p. 52) notes that under identical environmental conditions, the respiration rate of arctic plants is higher than that of plants from lower latitudes. Although the inference of climatic adaptation (temperature, length of daylight, etc.) seems reasonable in this case, the evidence cannot be said to constitute absolute proof.

The International Biological Program IBP is undertaking physiological studies of populations, particularly of primitive people such as the Australian aborigine. Another fruitful line of inquiry is a comparative study of groups of people living in the same city at the same economic level, whose parents immigrated from different climatic regions.

13/CLIMATIC CLASSIFICATION AND INDICES

13.1. HISTORICAL INTRODUCTION

The first climatic classifications were published in the 1840's (440). J. K. Klauprecht proposed a subdivision according to latitude and annual mean temperature, with a further separation according to whether a location was subject to marine, continental, or mountain influences. Independently in England, R. B. Hinds suggested 16 climatic zones, based on wetness, annual mean temperature, and annual temperature range.

A question that should be asked at the outset is why Klauprecht, Hinds, and all their successors, wished to classify the climate at all. As a partial answer, perhaps they were attempting to catalog the similarities in a search for order. Hinds seems to have recognized a general principle that vegetation zones are related to climate, although the correspondence is not exact by any means. In addition to the fact that many different nonlinear combinations of the meteorological variables may be limiting, the response of a plant community is not unique. Nevertheless, the principle is useful in a qualitative way for introductory discussions of world climate.

The number of climatic zones may be made so large that no anomalies remain. In practice, such precision would be impossible to attain, and even if approached, the resulting system would be too cumbersome to be of any value. An easy way to begin is to select a particular species, map its global distribution, and determine the ranges of the meteorological variables in the areas where the plant flourishes. This has some appeal. Hopefully, correct predictions can be made that the species may be

TABLE XL

MAIN GROUPS OF KÖPPEN'S CLIMATIC CLASSIFICATION[a]

| Climatic province | Symbol | Mean temperature (°C) | | Precipitation (cm) | |
		Coldest month	Warmest month	Main season	Amounts[c]
Rain forest	Af	>18		All year	Driest month >6
					Annual >100
Savanna	Aw	>18		Winter[b]	Driest month <6
					Annual >100
Steppe	BS			Winter[b]	Annual <2T
				Summer	Annual <2(T + 7)
				Even	Annual <2(T + 14)
Desert	BW			Winter[b]	Annual <T
				Even	Annual <(T + 7)
				Summer	Annual <(T + 14)
Moderate, oceanic	Cf	<18 >−3		Winter[b]	Annual >2T
Moderate, winter dry	Cw	<18 >−3		Even	Annual >2(T + 7)
Moderate, summer dry	Cs	<18 >−3		Summer	Annual >2(T + 14)
Oceanic boreal	Df	<−3		Even	Annual >2(T + 7)
Continental boreal	Dw	<−3		Summer	Annual >2(T + 14)
Tundra	Et		<10 >0		
Eternal frost	Ef		<0		

[a] Köppen (441).
[b] Astronomical winter.
[c] T is mean temperature in °C.

introduced successfully in other parts of the world having the same type of climate.

The continual refinement of climatic classifications was a nineteenth century pursuit, having its roots in the Darwinian theory of the survival of the fittest in each environment (436). It reached its ultimate in the work of Köppen (441), whose system has remained a standard for many years. Köppen's main groupings, given in Table XL, are based on mean temperature and precipitation. Köppen and others quite naturally were influenced in their choice of meteorological elements by what was actually available from a global network of stations. Only much later were attempts made to introduce more meaningful parameters such as those derived from the water budget or energy balance equations.

Although the Köppen classification is quite satisfactory for a general discussion of world climate, it does not meet the needs of many specialized consumers. The construction engineer requires a wind and snow-loading climatology while the airlines are interested in upper winds and surface landing conditions. Because the relevant meteorological elements and their class limits vary for each application, the objective of the classification must therefore be specified.

The suggestion has often been made that it is preferable to present individual charts for each meteorological element, permitting the user to draw his own conclusions. In fact, the World Meteorological Organization (158) emphasizes that an atlas must contain maps of individual elements. Even so, the cartographer has many options in his choice of indicators of frequency distributions. Consider, for example, precipitation. Suzuki (442) has attempted to develop a mesoscale classification for Japan based on this single element. He emphasizes that the rainfall distribution cannot be characterized by a single parameter but requires the arithmetic mean, standard deviation, coefficient of variation, lag correlation between rainfalls in consecutive years, persistencies of daily and monthly wet spells, ratio of June--July rainfall to annual total, intensity--duration indices, snowfall statistics, etc. Even for this single meteorological element, therefore, an investigator could spend a lifetime defining the climate of a small region.

13.2. AIR POLLUTION INDICES

Many people have searched for a single convenient pollution index. The various proposals may be classified according to the point of entry into the following chain:

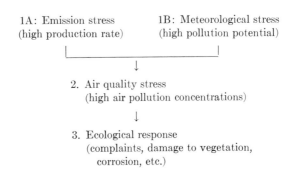

1A: Emission stress
(high production rate)

1B: Meteorological stress
(high pollution potential)

↓

2. Air quality stress
(high air pollution concentrations)

↓

3. Ecological response
(complaints, damage to vegetation,
corrosion, etc.)

Lag times between steps 1 and 2 in a city are usually short, of the order of hours, and are often predictable from a knowledge of horizontal and vertical ventilation rates. Lag times between steps 2 and 3 are variable and not always understood. A pollution index designed to reflect day-to-day variations should therefore be based on data of types 1A, 1B, or 2.

For studies of trends over decades, a suitable index may be some measure of response such as corrosion rates or the annual number of cases of emphysema. The urban pollution officer seeks simple indicators to test the effectiveness of his control programs in relation to expanding industrial development and population growth, and he wishes to remove the effect of meteorological "noise" by using a long averaging time. In many cities, however, there are no historical data on pollution concentrations or responses. Trends in some of the meteorological elements may therefore be examined. Relevant climatic indices include annual frequencies of smoke and haze, visibility distributions, hours of sunshine, and solar radiation flux. An interpretative difficulty is that most of the weather-observing stations are at airports, locations which are becoming more urban over the years; changes in visibility distributions thus are indicators of urban growth rather than of air quality trends within the city. Emission-stress data (step 1A) may also be used as pollution indices. Examples include the annual sales of various types of fuels and the number of registered motor vehicles.

To create a meaningful index of monthly or daily urban air quality stress (step 2), meteorological factors (step 1B) and emission rates (step 1A) must both be considered. There are annual and weekly cycles in the production of pollution. If these are known, meteorological stress indices (considered below) may be normalized by a multiplicative factor appropriate for the time of year and day of the week. If emissions for a weekday in January are 25% above the annual average, for example, the meteorological stress index is converted to a combined meteorological/emission index by simply multiplying by the factor 5/4. For study of a particular part of a city with sources mainly in one upwind sector, a separate adjustment is required for each wind direction.

Often no source inventory is available. Weisman *et al.* (443) have suggested the use of degree days as an indicator of emission strength for conventional types of pollution such as SO_2 and smoke. They have proposed a monthly pollution index PI_{mon}:

$$(13.1) \qquad PI_{mon} = 100 \left[\frac{C_{mon}}{C_{ann}} - \frac{\text{deg days}_{mon}}{\text{deg days}_{ann}} \right]$$

where

$\qquad C_{mon}$ = the monthly mean pollution concentration,

$\qquad C_{ann}$ = the annual mean pollution concentration,

deg days_{mon} = the number of degree days below 18.3°C (65°F) for the month,

deg days_{ann} = the annual number of degree days.

The method was tested in Hamilton, Ontario, using smoke density as an indicator. The index was highest in summer despite the fact that smoke concentrations were lowest at that time of year. Weisman *et al.* conclude that meteorological conditions create a high pollution potential more frequently in summer than in winter.

Equation (13.1) represents an attempt to infer a meteorologically dependent index (step 1B) from air quality data (step 2). In areas where emissions are as large in summer as in winter, the use of degree days would of course be misleading. Wherever possible, the preparation of source inventories is desirable.

Turning now to a direct meteorological stress index, Weedfall and Linsky (444) emphasize that three kinds of pollution are important:

(a) The products of incomplete combustion (SO_2, smoke, etc.).
(b) Photochemical pollution (oxidants, PAN, etc.).
(c) Fog-reactive gases (SO_2, etc.).

Therefore, three separate pollution potential indices may be required. Photochemical smog forms only in the presence of sunlight while SO_2 may accumulate at night or during overcast skies. In cities on the shores of the Great Lakes, oxidant levels occasionally become high in the summer; SO_2 and smoke peaks, on the other hand, are associated with the autumn and winter months.

Lowry and Reiquam (445) have proposed a pollution potential index for Salem, Oregon that includes the effect of persistence. A daily index DI is defined as follows:

$$(13.2) \qquad DI = 14 + (T_{9,03} - T_{s,03}) + (T_{9,15} - T_{s,15})$$

where T refers to temperature in °C, subscripts 9 and s refer to the 900-mb and surface-pressure levels, 03 and 15 refer to radiosonde flights at 0300 and 1500 PST. The factor 14 has been chosen subjectively so that the index DI is near zero on a day when there is not likely to be a net gain or loss of pollutants by vertical ventilation in the area.

To study episodes, the index DI is summed for consecutive days until there is an occurrence of one or more of the following conditions:

(a) A single negative value of DI (strong vertical ventilation).
(b) A large daily vector mean wind (strong horizontal ventilation).
(c) A daily precipitation total of 0.13 cm (0.05 in.) or greater (washout of pollution).

Lowry and Reiquam believe that this index is of value in anticipating pollution problems in areas where emissions are still low.

An entirely different approach is based on the qualitative relation that exists between the diffusion of pollution and the character of the wind-direction fluctuations of a sensitive vane. The Brookhaven classification (446) divides the "weather" into five characteristic patterns according to five types of recorder traces of wind direction. During sunny days with modest winds, for example, thermal convection causes fluctuations in the wind of more than 90°; with strong inversions on the other hand, turbulence almost completely disappears.

Useful as the Brookhaven classification has been for research investigations, its value is limited by a lack of historical data. Another choice is the Pasquill classification (16, p. 204), which requires hourly observations of surface wind speed, sunshine, and cloudiness, all of which are on punched cards for a large number of stations. Details are given in Table XLI. The weather has been divided into 6 classes, A–F, each of which

TABLE XLI
THE PASQUILL CLASSIFICATION[a]

Surface wind speed (m/sec)	Insolation			Night	
	Strong	Moderate	Slight	Thinly overcast or $\geq$ 4/8 low cloud	$\leq$ 3/8 cloud
<2	A	A–B	B	—	—
2–3	A–B	B	C	E	F
3–5	B	B–C	C	D	E
5–6	C	C–D	D	D	D
>6	C	D	D	D	D

[a] Pasquill (16).

has characteristic turbulence and diffusion regimes. The climatological frequency of occurrence of each class has been determined for a number of stations in England (447) and elsewhere.

Pasquill's classification has a much wider applicability than is generally recognized. Because classes A–F describe the diffusive capacity of the atmosphere in a simple but physically satisfying way, they may be used, for example, in water balance investigations. Instead of calculating monthly evapotranspiration from monthly mean values of the appropriate meteorological variables, prestratification according to a Pasquill-type classification is recommended.

13.3. WATER BUDGET AND SOIL MOISTURE INDICES

There has been a continuing search for magic numbers in studies of the hydrologic cycle and soil moisture. Most of the indices recognize the importance of precipitation P as well as of evapotranspiration ET, but they differ widely in detail.

Indices may be required for the following three reasons:

(a) For stations or watersheds with substantial instrumentation, there may be a desire to parameterize the relevant factors, or at least to reduce the number of variables. Examples include the use of a single meteorological index for correlation with crop yield, and the search for an empirical form of the *runoff ratio* R/P, where R and P are runoff and precipitation, respectively.

(b) It may be useful to compare locations, given only the simplest kinds of climatological data. If one area has twice the rainfall of another but also twice the evapotranspiration, and both have moisture deficiencies, how are they to be compared?

(c) It is often useful to rank different years at the same location. Was a particular summer dry, and if so, was that due to below-average rainfall or above-average transpiration? Other equally difficult questions could be posed.

A number of indices are derived from the water budget equation (8.14). On an annual basis the storage term can be neglected usually and

$$(13.3) \qquad P = \mathrm{ET} + R$$

As noted by Sellers (448, p. 85), this may be rewritten in the form,

$$(13.4) \qquad R/P = 1 - \mathrm{ET}/P$$

Thus the *runoff ratio* R/P depends on another index ET/P. Empirical forms have been suggested for each of these parameters, but sometimes without realization that they must be connected by Eq. (13.4). Sellers (448, p. 89) suggests that for the United States,

(13.5) $R/P = cP$

where c is constant for any particular watershed but variable from place to place; the value of c must be estimated experimentally.

Using annual averages, another index has been used by Budyko (449), PET/P, which is called the *radiational index of dryness*. Because PET is strongly correlated with Q_N on this time scale, the index is also given as Q_N/LP, where L is the latent heat of condensation. This leads to a geobotanical classification:

$$Q_N/LP < 0.35 \qquad \text{tundra}$$
$$0.35 < Q_N/LP < 1.1 \qquad \text{forest}$$
$$1.1 < Q_N/LP < 2.3 \qquad \text{steppe}$$
$$2.3 < Q_N/LP < 3.4 \qquad \text{semiarid}$$
$$3.4 < Q_N/LP \qquad \text{desert}$$

Lettau (450) has noted that the water budget and heat balance equations may be combined to yield,

(13.6) $(1 + Q_H/Q_E) (1 - R/P) = Q_N/LP$

Thus, there is an important connection amongst the runoff ratio, the radiational index of dryness and the annual value of the Bowen ratio Q_H/Q_E.

Thornthwaite (368, p. 69) has emphasized that PET/P is a better indicator of aridity than is ET/P. Thornthwaite and Mather (451) have defined an annual *moisture index* I_m:

(13.7) $I_m = 100 \dfrac{(\text{moisture surplus} - \text{moisture deficit})}{\text{moisture need}}$

$$= 100 \left[\frac{R - (\text{PET} - \text{ET})}{\text{PET}} \right]$$

But $R = P - \text{ET}$. Thus,

(13.8) $I_m = 100[P/\text{PET} - 1]$

A value of $I_m = 0$ occurs when water supply is equal to water need. Positive values indicate that there is a surplus of precipitation. Subrahmanyam and Sastri (452) have used I_m as a drought index. Severe and disastrous drought years are defined as those in which I_m is more than one and two standard deviations, respectively, below its median value for a given location.

The critical question in this and other similar work is whether ET/R or PET/R is a better indicator for the particular objective of the study. The burden is on the investigator to explain the physical basis of his index and to demonstrate its value in comparative experiments at locations equipped with instruments for measuring *all* rather than a few of the relevant variables.

Gary (453, p. 62) has used a quantity called *effective precipitation:*

$$(13.9) \qquad \text{effective precipitation} = \frac{\text{precipitation}}{\text{wet bulb depression}}$$

where annual or monthly averages are implied. Wet bulb depression increases with both rising temperature and with decreasing relative humidity; thus, it is approximately proportional to PET. An atmometer too (Section 11.2) is essentially an indicator of wet bulb depression. Equation (13.9) is interpreted to mean that as the rate of evaporation increases, less and less of the precipitation is available for soil moisture recharge.

13.4. CONTINENTALITY INDICES

The *continentality index* C has been defined as follows:

$$(13.10) \qquad C = c_1[(\bar{T}_{\max} - \bar{T}_{\min})/\sin \phi] - c_2$$

where

$\bar{T}_{\max}, \bar{T}_{\min}$ = mean maximum temperature of the hottest and the mean minimum temperature of the coldest month,

ϕ = the latitude,

c_1, c_2 = constants determined arbitrarily by setting $C = 0$ at a maritime location,

$C = 100$ at a very dry station.

This index has been used, for example, by deBrichambaut and Wallén (454, p. 19) in a study of the Near East and by Kopec (455) in an investigation of the Great Lakes watershed. Regional maps of the index are of qualitative interest, but of little quantitative value.

Polowchak and Panofsky (456) have proposed a continentality index which is based on spectra of temperature. To remove the annual cycle, the time series which they employ are the departures of the daily mean

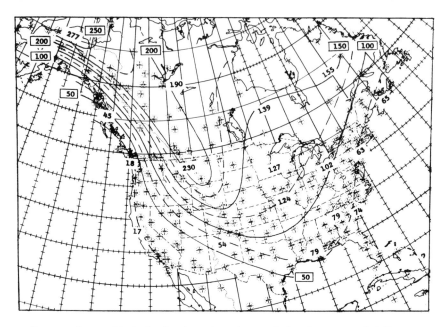

FIG. 95. Geographical distribution of variance of the departure of daily mean temperature in winter from the climatological normal for the day (456).

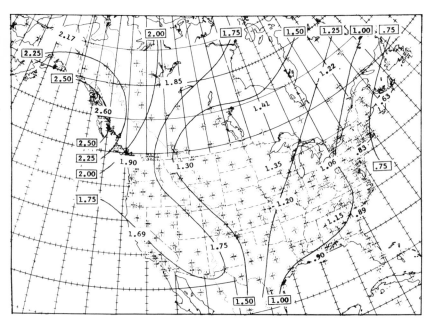

FIG. 96. Geographical distribution of winter rhythm index as defined by Eq. (13.11) (456).

temperatures from the climatological normals for each day. Two properties of the spectra are used as indices:

(a) The total variance (Fig. 95), which is a measure of interdiurnal variability.

(b) A *rhythm index* RI (Fig. 96), defined as follows:

$$(13.11) \qquad \text{RI} = \frac{2(S_3 + S_4)}{S_7 + S_8 + S_9 + S_{10}}$$

where S is the spectral estimate at the frequency (cycles per 48 days) indicated by its subscript. The index is unity for white noise (all S_i equal); a smaller value indicates that high frequencies predominate and vice versa. Some results (Figs. 95–96) for the winter period (November 1–February 28) are physically reasonable and suggest that these are indeed useful continentality indices.

13.5. AGRICULTURAL INDICES

In addition to moisture stress, discussed in Section 13.3, many factors contribute to agricultural yield. The WMO *Guide to Agricultural Meteorological Practices* (457) lists the following variables:

(a) Elements important to vegetative growth:

(1) Moisture stress (including an excess of water).
(2) Radiation and temperature.
(3) Duration of the frost-free period.

(b) Elements important to the development of successive phases in plant life:

(1) Day length.
(2) Annual variation in radiation and temperature.
(3) Daily temperature range.
(4) Duration of the frost-free period.
(5) Durations of rainy and dry seasons, where such exist.

Attempts to combine all these variables into one or two indices must necessarily be incomplete. Although human factors (technology) and the occurrence of rare events such as hail storms add further dimensions to the problem, there has been modest success in particular cases.

First, the principal environmental stresses in each region should be identified. In the tropics, for example, photoperiod and frost may quickly be eliminated from an initial list of possible stresses. Despite this simplification however, separate indices may be required for different crops, and for irrigated as well as natural conditions.

Burgos (458) suggests that three separate approaches should be used in the search for relations between global distributions of a particuar species and cimate:

(a) Evaluation of the agroclimate of the native region of the species. The plant has adapted itself there not only to average conditions but also to those extremes that occur perhaps only once a century.

(b) Evaluation of the agroclimate of other regions where the species is grown.

(c) Evaluation of the agroclimate of regions where experience has shown that the species cannot be grown.

Because soil conditions must also be considered, the climatologist should never attempt this kind of investigation alone.

Burgos has used the method in a study of seed potato. From an analysis of world distributions and associated climates, he infers that five factors are important: the water balance during the growing season, the mean temperature at the date of sowing, the mean maximum temperature of the warmest month during the growing season, the temperature amplitude during the warmest month, and the length of the growing season (to first killing frost). A range of values for each factor is established, delineating the climatic zones where seed potatoes are likely to flourish.

Burgos bases his studies on meteorological ranges rather than on means. Haggett (162), however, suggests that ranges are often poorly defined and should be replaced by central tendencies, i.e., by the mean and the standard deviation (or geometric standard deviation) of each meteorological element or derived index, for all locations where the species is growing successfully.

Finally, a climatic index for comparing conditions at one location year by year is given by agricultural yields. After removing the upward trend caused by improving technology, the remaining variability is a direct indicator of year-to-year variations in "weather." Therefore, a long time-series of yields may be a measure of atmospheric stresses. For regional comparisons of annual yields, Hustich (459) recommends use of the coefficient of variation [Eq. (6.2)], which he calls a *climatic crop-hazard indicator*. Hustich reports that the index is larger in northern than in southern Finland for yields of rye and for the growth of pine and spruce, indicating latitudinal control of climatic variability.

Simple agricultural indices are sometimes effective. The emergence of grasshoppers occurs quite suddenly and is closely related to accumulated growing degree days during dry summers. A reasonably precise forecast of the principal emergence date can therefore be made a week in advance by extrapolation of the observed cumulative total degree days. Similarly, soil temperature in spring is a reasonably conservative element and a useful predictor of a number of phenological events.

13.6. HUMAN COMFORT INDICES

Man responds nonlinearly to meteorological stress, as does every other living organism. The weather can be too warm or too cold for human comfort, just as rainfall may be excessive or deficient for optimum plant growth. Admittedly the comfort zone is rather uncertain, varying with age, state of health, type of clothing, and other factors. Nevertheless, the range is usually sufficiently broad to be useful, at least qualitatively, in developing indices.

As in hydrology and agriculture, there is frequently a desire to represent human climate by a single index, or to rank various locations according to cumulative environmental stress; this may be an annual sum or a summer "resort" index. Landsberg (258, pp. 130–133) has emphasized the importance of developing also a "retirement climatology" for elderly people.

A large number of human comfort indices have been suggested. Some of these were described in Chapter 9. Terjung (460), for example, has used effective temperature as a basis for developing an annual physioclimate for the United States. He defines an annual cumulative stress CS:

$$(13.12) \qquad \text{CS} = \sum_{1}^{365} (D^2 + N^2)$$

where D, N are the day and night stresses, respectively. Summation is over the year. The quantities D and N are squared because most human responses seem to fit such a nonlinear relation.

Terjung obtained values of D and N by subdividing a psychrometric chart into 11 areas, based on published data for physiological response. Each area was assigned a weight, ranging from $+4$ to -6:

+4 extremely hot (ET $> 30°$C: see Fig. 66, p. 196)
+3 sultry
+2 hot
+1 warm
 0 normal
−1 cool
−2 keen
−3 cold (ET $< 1.7°$C)

−4 very cold (ET < −10°C)
−5 extremely cold (ET < −20°C)
−6 ultracold (ET < −40°C)

Values of D and N are obtained from daily observations of maximum and minimum temperature and relative humidities. To simplify computation, Terjung used monthly mean values of these quantities. The resulting cumulative stress for the United States is displayed in Fig. 97. The index decreases southward, although there is an increase again in some equatorial regions; Port Sudan on the Red Sea has a CS of 124, for example. If such weighting is accepted, then winter cold in the north produces more stress than summer heat and humidity in the south. Values of CS for Minnesota are due almost entirely to negative weights while the reverse is true for Florida. Terjung suggests that the American middle West has a very "intemperate" climate. It is to be noted, however, that the use of monthly mean maximum and minimum temperatures results in lower values of CS than if daily observations were used; the effect on the isopleth gradients in Fig. 97 remains to be investigated.

Bailey (461) has defined a *temperateness index M*, and his map of the United States (Fig. 98) resembles in many ways that of Fig. 97. Bailey begins by assuming that a mean annual temperature of 14°C is a reasonable central value for human comfort and for biological activity. He then calculates the second moment of departures of hourly temperatures from a reference value of 14°C over a one-year period, i.e., $\langle (T - 14)^2 \rangle_{av}$. The temperateness index M is defined as

(13.13) $$M = 109 - 30 \log \langle (T - 14)^2 \rangle_{av}$$

The constants have been selected so that M is near zero in Antarctica and near 100 in some tropical uplands. The logarithmic term permits the very large range in the second moment to be compressed into a 0–100 scale. Bailey admits that his index "will repel those with discriminating statistical tastes" but he believes that his empirical expression may have value.

Hounam (462) has used the relative strain index, Eq. (9.16), to produce a heat stress climatology for Australia, based on midafternoon synoptic weather observations. Figure 99 shows the average number of days per year for which the relative strain equaled or exceeded 0.3. This type of map has many applications.

Foord (463) and others have employed effective temperature (directly) as a climatic index. Foord, for example, has determined the frequency of uncomfortably warm days in London, England.

Among the many other proposals for a single index, Gregorczuk (464) has used *enthalpy*, or total heat content of the atmosphere exceeding arbitrarily selected zero points, −18°C for dry air and 0°C for water

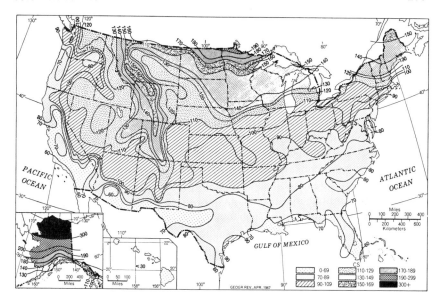

Fig. 97. Annual cumulative stress CS, as defined by Eq. (13.12), for the United States (460). [Reprinted from the *Geographical Review* (Vol. 54, 1964), copyrighted by the American Geographical Society of New York.]

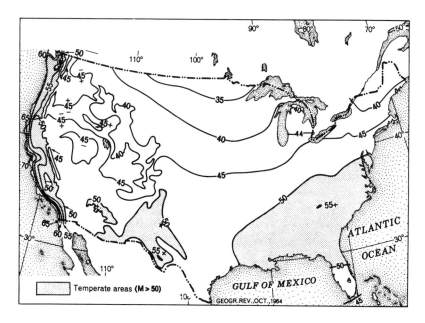

Fig. 98. Temperateness index M, as defined by Eq. (13.13), for the United States (461). [Reprinted from the *Geographical Review* (Vol. 54, 1964). Copyrighted by the American Geographical Society of New York.]

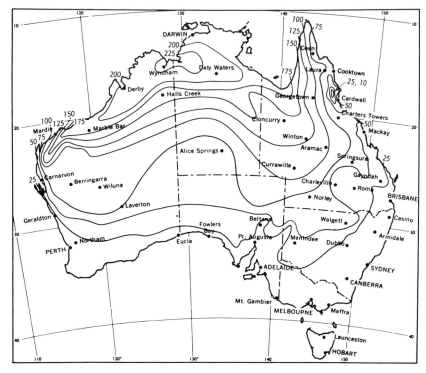

FIG. 99. Average number of days per year when relative strain index equaled or exceeded 0.3 at 1500 hr over Australia (based on data for the period 1957–1966) (462).

vapor. He suggests that one advantage of this index is its dimensions (kcal/kg), which are consistent with those of metabolic units, kcal/kg-hr. Böer (163) has included a formula for enthalpy i.

$$(13.14) \qquad\qquad i = c_\text{p}(T_\text{w} + 1.555p/e_\text{s})$$

where

$\qquad T_\text{w}$ = wet-bulb temperature in degrees Celsius,
$\qquad p$ = atmospheric pressure in millimeters of mercury,
$\qquad e_\text{s}$ = saturated water vapor pressure at the air temperature.

Gregorczuk has used Eq. (13.14) to prepare global maps of air enthalpy.

Maunder (465) suggests that other factors besides temperature and humidity are important in a human climatic classification. He notes that the New Zealand cities of Hamilton and Wellington are in the same climatic zone according to both Köppen and Thornthwaite; yet Wellington averages 151 days a year with wind gusts of at least 17.9 m/sec (40 mph) and has no "screen" frosts while Hamilton has only 23 days with strong gusts but 29 days with frosts. These differences are important to the layman. Maunder suggests an index for New Zealand that includes rainfall, sunshine, temperature, humidity, and wind, subjectively weighted according to the proportions

$$R:S:T:H:W = 8:7:6:5:4$$

Further subdivisions are made in each class to arrive at an index:

$$(13.15) \quad X = (3P_1 + 3P_2 + 2P_3) + (4S_1 + 3S_2) \\ + (2T_1 + T_2 + T_3 + T_4 + T_5) + (5H_1) + (2W_1 + 2W_2)$$

where

X = human climatic index,
P_1 = mean annual rainfall,
P_2 = mean annual duration of rain,
P_3 = percentage of rainfall 9 PM–9 AM,
S_1 = mean annual duration of bright sunshine,
S_2 = mean winter duration of bright sunshine,
T_1 = mean annual number of degree days,
T_2 = mean number of days with screen frost per year,
T_3 = mean daily maximum temperature of coldest month,
T_4 = mean annual maximum temperature,
T_5 = mean number of days with ground frost per year,
H_1 = "humidity index,"
W_1 = mean number of days with wind gusts 17.9 m/sec and over,
W_2 = mean number of days with wind gusts 26.8 m/sec (60 mph) and over.

Each of the 13 variables is given a rating from 1 (very favorable) to 5 (very unfavorable). For example, W_1 ratings 1, 2, 3, 4, 5 correspond to annual numbers of gusty days of 0–14, 15–44, 45–89, 90–149, and 150–224, respectively. The resulting values of X for 22 stations in New Zealand range from 60 at Nelson (most favorable) to 101 at Invercargill (most unfavorable). The smallest and largest possible values of X are 30 and 150. Maunder's approach has much to commend it in principle, although many arbitrary decisions had to be made in selecting the variables and their weights. In some other areas, for example, the frequency of snowstorms is important.

Davis (466) has described a number of summer weather indices that have been proposed for Britain. He then suggests

$$(13.16) \qquad I = 10T + 20S - 7P$$

where for the three months June, July, August,

I = human climate index,
T = the mean daily temperature in degrees Fahrenheit,
S = daily mean sunshine in hours,
R = the total rainfall in inches.

In conclusion, the very fact that so many indices have been suggested indicates the need for further research, particularly into the question of population responses to multiple environmental stresses. That the search is worthwhile has been emphasized by Prohaska (467). With increasing global population and attendant food and water consumption, ecosystems are being examined in more and more detail. To reduce the number of variables, there is a need for simple atmospheric indicators that integrate the effects of several environmental stresses. Undoubtedly, separate indices are required for separate ecosystems but this should not deter the investigator in his search.

13.7. CLIMATONOMY

Lettau (See Ref. 450, for example) has proposed a new word *climatonomy*, which he defines as the study of methods of describing and predicting climatic responses to various forcing functions, using very general physical models. An appropriate climatonomy would permit mesoscale mapping of the radiation balance of rugged terrain, given the balance at one point and the spatial distribution of slope angles, aspects, and ground cover.

Lettau (450) has discussed an evapotranspiration climatonomy for continental areas. The forcing functions are solar radiation and precipitation, and the response is monthly evapotranspiration. The governing relations are the surface heat balance and water budget equations. These are transformed to dimensionless form, Eq. (13.6), for example. In order to obtain a closed set of equations, a few additional assumptions are required, and dimensionless constants must be introduced. These constants, however, may often be interpreted physically and may be used as climatic indices.

As an example, Lettau separates monthly evapotranspiration ET into ET′ and ET″, where ET′ is moisture with a relatively short "turnaround" time, i.e., the water has moved through the precipitation–evaporation hydrologic cycle within the given month, and ET″ is moisture with a storage time in the soil of greater than one month. Lettau then postulates that ET′ varies linearly as the monthly product of precipitation P and absorbed insolation Q. The quantity Q is made nondimensional by expressing it as a fraction of its annual local mean $\bar{Q}$. Finally,

$$(13.17) \qquad\qquad ET'/P = cQ/\bar{Q}$$

The nondimensional parameter c, called *evaporivity* by Lettau, varies in space and perhaps in time but it may "summarize" many of the hydrologic features of a region or watershed.

In this and other climatonomies, the study of characteristic times is often informative. These parameters may be derived from a dimensional analysis or may become apparent when a governing equation is made nondimensional. The *residence* time is the time required for N tagged molecules in a volume at some given time to be reduced to $0.368N$ molecules (see Eq. 2.13). The *turnover time* is the ratio of biomass to productivity (see Section 9.5); alternatively, for the atmosphere, the turnover time is the ratio of the mass of a particular substance in a given volume to its net loss per unit time out of the volume. As an illustration, if the global annual assimilation of CO_2 by vegetation is $0.03N$, where N is the atmospheric mass of CO_2, the turnover time is $N/0.03N$, i.e., about 35 years. These characteristic times may be useful biometeorological indices.

14/ENGINEERING AND ECONOMIC APPLICATIONS

14.1. IMPACT OF WEATHER AND CLIMATE ON HUMAN ACTIVITIES

Day-to-day weather has a large economic impact (468). Although each individual responds differently, large populations reveal discernible group behavior. Sales of air-conditioning equipment increase during the first heat wave of the season, as do purchases of overshoes in the first winter snowstorm; even church attendance is weather dependent (469). Advertising agencies and retail stores are aware of the effect of weather on buyer resistance, although few quantitative studies have been published. Daylight illumination has an influence on urban hydroelectric consumption while dry cleaning and laundry establishments are busy during periods of poor air quality.

Not only the weather but also the forecast is important, whether the prediction proves to be correct or not. Weekend travel to the country is reduced if stormy conditions are forecast, and the sale of tickets to outdoor events is lowered if rain is predicted. Probability forecasts are now being issued daily to the general public in the United States. Many weather-sensitive operations (business, agriculture, transportation, etc.) can use the information to make regular economic decisions; the losses incurred when rain falls must be weighed against the cost of taking evasive action (see Section 14.3).

National weather services and private consultants are asked frequently for climatic information or analysis. The design of a large dam or a tall building requires a detailed climatological study, and a decision must be made sometimes with incomplete data. There is also reason to believe that sociological patterns are dependent on climate, or at least on human attitudes towards climate. Even on the microscale, climate is a locational factor; stores on the sunny side of the street attract more customers than do stores in the shade (except in very hot weather).

14.2. ENGINEERING METEOROLOGY

Meteorologists study the relationships that exist in time and space among the meteorological variables. Simplified physical models are often used; the relation between vertical flux and gradient over a uniform surface (Section 2.5) is an example. Climatologists examine the same space scales but over longer time periods.

Because the surface of the earth is so irregular, many of the physical models are not always directly applicable to real situations such as the wind flow within a built-up area in a city or the evapotranspiration from a complex watershed. There is need, therefore, for the meteorological engineer. This problem-solver seeks order-of-magnitude solutions, based on a knowledge of the local topography and on an understanding of the existing physical models, their underlying assumptions and their limitations.

Page (470) has considered three important aspects of this general question.

(a) *Time:* Time is a critical factor in the decision-making process. The engineer must make thousands of individual decisions in the construction of a single building and he cannot wait for "exact solutions."

(b) *Safety:* Infinite safety is achieved only at infinite cost. A reasonable compromise must therefore be made, although the decision should always be based on rational scientific evidence. There must be a continuing program of case-history studies, determining why some wrong decisions were made on previous occasions; this might be called the method of successive approximations.

(c) *Feedback:* When an engineer rejects meteorological information, he does so for one of three reasons, namely, because the information is irrelevant, inapplicable, or incomprehensible. Each of these adjectives implies a lack of communication between the two professions.

14.3. THE COST/LOSS RATIO

The cost/loss ratio C/L is used widely in meteorological economic studies. Let

C = the cost of protection against an adverse event,

L = the loss if protective measures are not taken and the event occurs,

p = the forecast probability that the event occurs.

Then protective action should be taken if $p > C/L$. A state of indifference $(p = p_0)$ exists if $p_0 = C/L$. For example, suppose that the forecast probability of rain is 0.1 and that the cost of protection against rain is $1000. The critical value for L is therefore $10,000. Presumably, evasive measures should be initiated if a loss of more than $10,000 would otherwise ensue. Reversing the logic, if C, L are given as $1000, $10,000, respectively, then $p_0 = 0.1$. On any day when the probability of rain is greater than 0.1, appropriate protective action should be taken.

There are many reasons why the decision process is rarely this simple. Although a large company may be prepared to accept a $20,000 loss on a single occasion (provided that the strategy is successful on the average over many repetitions), a small company may be placed in a serious financial position by such a loss. Thus, the probability p_0 for indifference may be $p_0 = 1$, that is, continuous protection. The operator must make a subjective evaluation of his particular p_0, a process called *maximization of utility*. In some cases he may be willing to gamble if the rewards are sufficiently high.

Suppose that a highway engineer has not completed his contract on time, resulting in a $20,000/day penalty for further delays. Suppose also that the cost of protection against rain is $1000, and that the loss would be $15,000 if rain were to fall at a time when the fresh concrete was not covered. If the probability of rain is 0.05, the operator is likely to protect, despite the fact that $p < C/L$. As another example, let p be the probability that airport weather will be "above limits" at the scheduled time of arrival of an aircraft. Let P be the net profit from a successful flight, L_1 be the net cost of overhead resulting from cancellation, and L_2 be the loss caused by diversion of the aircraft to another airport. Gringorten (471) notes that the indifferent probability p_0 is given by,

$$(14.1) \qquad\qquad L_1 = (1 - p_0)L_2 - p_0 P$$

Rearranging terms,

$$(14.2) \qquad\qquad p_0 = (L_2 - L_1)(L_2 + P)^{-1}$$

It can be seen that if P is relatively very large or if L_1 and L_2 are approximately equal, the indifferent probability p_0 is very small; the dispatcher should then try to maintain his schedule in adverse weather conditions.

In this and many other examples, the utility of the decision is significant. Passenger inconvenience caused by cancellation of the flight must be weighed against that resulting from diversion of the aircraft to an alternate airport. Other factors that are sometimes important include comfort, safety, and "moral worth." Shorr (472) defines a cost/loss utility ratio $U_1(C)/U_2(L)$, where U is a *utility function* that varies with C, L, the decision-maker, and the nature of the operation. Using an intuitive argument, Shorr suggests that

$$(14.3) \qquad \frac{U_1(C)}{U_2(L)} = \frac{C}{L}\left(\frac{1 - L/R}{1 - C/R}\right)$$

where R is the ruin, bankruptcy, or customer utter dissatisfaction cost, which must be specified by each decision maker. In many cases, $C \ll R$. Then

$$(14.4) \qquad U_1(C)/U_2(L) = (C/L)(1 - L/R)$$

For C, L values of $1000, $10,000, respectively, the ratio C/L, or p_0, is 0.1. If the bankruptcy value is $20,000, however, the cost/loss utility ratio obtained from Eq. (14.4) is 0.05.

A central problem in meteorological cost/loss studies is the assessment of the probability p of a weather event. Quite frequently the climatological probability is too small to be of much operational value or it may not be known for a specific location some distance from a regular observing station. The alternative is to use 36-hr weather forecasts, normally of greater accuracy than climatological or persistence predictions. Thompson (473) has given a simple example. An industry customarily brought equipment indoors nightly during the winter, at a cost of $300 per night. This provided protection against the risk of subfreezing temperatures, the occurrence of which would result in a loss of about $2000. Over a 90-day period, the total cost of protection was therefore $90 \times 300, or $27,000. The reliability of minimum temperature predictions is given in Table XLII. If these forecasts had been used, protective measures would have been taken on 57 cases instead of the

TABLE XLII

THE RELIABILITY OF A SAMPLE OF 90 FORECASTS OF MINIMUM TEMPERATURE[a]

Observed	Predicted		Total
	Freezing	No freezing	
Freezing	51	6	57
No freezing	6	27	33
Total	57	33	90

[a] Thompson (473).

entire sample of 90; however, there would have been 6 failures, i.e., 6 nights of frost when the equipment would have been left outdoors. The total cost would therefore have been

$$57 \times \$300 + 6 \times \$2000 = \$29,100$$

This illustrates an important point that has been emphasized by several authors. Although the forecaster showed considerable skill (only 12 failures in a sample of 90), his predictions were of no economic value in this particular operation.

The cost/loss ratio for this example was 300/2000 or 0.15. Over the same 90-day period the meteorologist was asked to make a daily subjective estimate of the probability of overnight freezing weather. The

TABLE XLIII

The Reliability of a Sample of 90 Forecasts with a Subjective Probability Greater Than 0.15 of Freezing Weather[a]

	Predicted		
Observed	Freezing	No freezing	Total
Freezing	56	1	57
No freezing	14	19	33
Total	70	20	90

[a] Thompson (473).

results are given in Table XLIII, using a probability of 0.15 as a criterion for protection. The total cost now becomes

$$70 \times \$300 + \$2000 = \$23,000$$

This is a saving of $4000 when compared with the cost of nightly protection.

The potential economic value of the method is not to be doubted but subjective estimates of weather probabilities and confidence limits are difficult to assess. The value of p varies with the experience of the forecaster, the meteorological element, the location, the season, and even the year. In addition, if there is a verification system, the forecaster is likely to choose the alternatives that yield the highest skill score. Although there is a trend towards objective methods of obtaining weather probabilities, a further problem arises, namely that the confidence limits often seem to vary with time.

This has been a superficial overview of a very complex topic. Borgman (474) and Epstein (475) suggest application of the theory of games in which "our opponent (nature) is unaware of the rules of the games (i.e., the system of awards, the utilities of the outcomes) but has chosen a strategy nevertheless and will adhere to that strategy, in the future. We do not know what his strategy is, but must estimate that strategy on the basis of his past performance" (475).

14.4. WEATHER SERVICES AND THE NATIONAL ECONOMY

Some people grumble that the layman has lost his ability to predict the weather from "signs in the sky" because of the widespread dissemination of forecasts on radio and television. Nevertheless, there can be no doubt that such information is valuable to the national economy. The loss of life in hurricanes was much greater in the nineteenth century when there were no storm warnings than it is today. An example has been given by Arakawa (476), that of a typhoon that struck Hiroshima when communications were still blacked out 42 days after the nuclear attack; casualties were more numerous in Hiroshima than on the island of Kyushu where the storm struck with greater force but with fore-warning.

Mason (477) has indicated in general terms the plans in the United Kingdom for the development of cost–benefit studies of weather services. In the United States too, analyses of this type are underway. Mason suggests that if daily forecasts and climatological summaries make a difference of 1% in agricultural productivity, the annual financial return is about £20 million in Britain whereas the cost of the meteorological services to that sector of the national economy is only about £50,000.

Economists and meteorologists must work together in such endeavors; neither can produce sensible results alone. An illustrative example is the problem of estimating the economic value of the establishment of an additional synoptic or radiosonde station. A possible approach is through computer simulation: a large number of numerical forecasts may be recreated using synoptic data but with the deletion of one strategically located station. Controlled verification procedures are required but the necessary methodology is being perfected. Alternatively, spatial correlations or eigenvector patterns (Section 6.5) may assist in determining the contribution of a particular station to forecast accuracy. There remains the problem of translating improved forecast reliability into economic terms. Gaudin and Kagan (478) have considered this question. They suggest that undoubtedly there is an optimum separation distance between weather stations, with diminishing economic returns for closer spacing. Because the cost of an observing station varies with location,

being largest in the arctic and on weather ships, and because user
requirements are difficult to evaluate, no general solution can yet be
given. Gaudin and Kagan emphasize, however, that cost-benefit analyses
should not overlook the value of expenditures on supplementary stations
for scientific research, the economic benefits of which may not become
apparent for a decade or so.

In the case of crop insurance, rates are often established empirically
from records of claims in previous years. The hail insurance rates in
Saskatchewan are shown in Fig. 100 (385); the isolines are based on

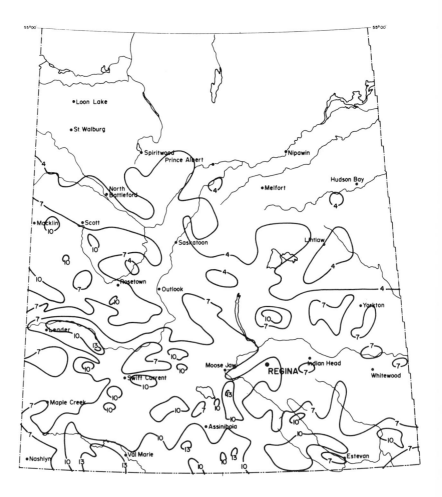

Fig. 100. Basic hail insurance rates (percent) in Saskatchewan. Higher values
indicate greater chances of hail (385).

actual payments to farmers over the period 1913–1966. Despite considerable scatter, there is a tendency for higher values in the south and southwest parts of the province. There is no way of knowing whether an increase in the length of record would change the patterns greatly, although, for example, the existence of "tornado alleys" in the midwestern United States seems to be well established.

Alternatively, as has been tried in at least one Central American country, national crop insurance rates may be derived from climatic data; a farmer may, of course, grow sensitive fruits in areas where there is a risk of frost or drought but then he may not qualify for compensation for crop losses. The climatic zones are determined from simple measurements of daily rainfall and maximum and minimum temperature, and a safety factor may be included to avoid local microscale anomalies. This procedure. encourages farmers to grow crops most suited to their environment.

There have been many studies of the economic losses due to pollution. As early as 1956, Scorer (21) attempted to introduce meteorology into the calculations, and he concluded that domestic sources of pollution in Britain contributed far more to property damage than did industry. Recent computer modeling of pollution transport in urban areas now provides an opportunity for simulation studies (323). Having successfully reproduced the existing air quality patterns in a city from a source-strength inventory and from some simplified diffusion equations, the investigator may apportion the contributions from various types of emitters and he may consider the effect of industrial development, population expansion, and/or a change in the type of fuel.

Because the agricultural industry is particularly sensitive to weather, economic studies can be of special value in that case. Some examples have been given by Edwards (479) and by de Wiljes and Zaat (480). Edwards suggests that it is desirable to undertake a cost-benefit analysis, based on the climatological probabilities of frost occurrence after various given dates in spring, before deciding whether to purchase frost-protection equipment. De Wiljes and Zaat indicate that in the Netherlands, harvest labor costs should be balanced against the climatological number of "agriculturally dry" days in the selection of crops that are likely to maximize profits.

Finally, weather events may set economic chains into motion. Smith (481) mentions that a recent drought in Argentina reduced the export of beef to Europe. The resulting increase in the price of fat cows caused a drop in the size of the British milk herd and a decrease in milk production.

Appendix/PROBLEMS

1. If an object is touched and feels neither hot nor cold, what can be said about its surface temperature?

2. If the autocovariance of a series of monthly mean values shows a marked peak at a lag of 26 months, what may be inferred, if anything, about its spectrum?

3. Describe the climate of your region if the earth were to rotate in the opposite direction.

4. Describe the climate of your region if:
   (a) the day were 12 hr in length,
   (b) the day were 48 hr in length,
   (c) the year were 6 months in length,
   (d) the year were 24 months in length.

5. If the Arctic Ocean were melted by some man-made method, would it freeze over again? [See, for example, "Can we control the arctic climate?" P. M. Borisov, 1967, *Priroda* **12**, 63–73; Engl. transl. Def. Res. Board, Ottawa, T 498 R (May, 1968).]

6. Why is monthly mean temperature more highly correlated with potential evapotranspiration than with actual evapotranspiration?

7. All other factors being equal, would evapotranspiration be the same at sea level as at an elevation of 2000 m?

8. Why is white clothing worn in the tropics?

9. Prepare a climatic classification for ski resorts. How is this modified if the areas are also to be used for summer recreation?

10. A large semiarid region is irrigated to the extent that evapotranspiration is at the potential rate. Will the net radiation flux remain the same?

11. Compare the heat balance of a man in a steam bath and in a sauna. [See, for example, "Physiological effects of extreme heat as studied in the Finnish sauna bath," J. Hasan, M. Karvonen, and P. Phronen, *Am. J. Phys. Med.* **46**, 1226–1246 (1967).]

12. Could a lake be used as a large rain gauge? [See, for example, "A comparison of rainfall data obtained from rain-gauge measurements and changes in lake levels," G. E. Harbeck, Jr. and E. W. Coffay, *Bull. Am. Meteorol. Soc.* **40**, 348–351 (1959).]

13. In an urban ragweed control program, will there be any significant reduction in pollen counts in the downtown area if only the land within the city limits is cleared of ragweed?

14. A cloud of insects often seems to hover over a person outdoors on a summer evening when winds are light. Why?

15. Describe the area–depth curve used in precipitation studies, and compare the concept with that of dosage–area in Section 2.2 [see, for example, "Area–depth curves—a useful tool in weather modification experiments," F. A. Huff, *J. Appl. Meteorol.* **7**, 940–943 (1968).]

16. In the measurement of rainfall, radiation, and wind over slopes, should the sensors be leveled according to the true vertical or should they be tilted?

17. Geiger ("Climate near the Ground," p. 106. Harvard Univ. Press, Cambridge, Massachusetts, 1965) suggests that although a vertical gradient of vapor pressure or specific humidity implies a flux of water vapor, this is not necessarily so in the case of a vertical gradient of absolute humidity (the weight of water vapor per unit volume of dry air). Why?

18. Compare the methodologies outlined in Sections 4.4 and 4.6.

19. (a) When there are waves on a lake, the area of the interface is greater than when the lake is smooth. Will this increase the evaporation rate? [See, for example, "Surface area of a wave-disturbed sea," A. S. Monin, *Izv. Atm. Ocean Phys.* **3**, 667–670 (1968); Engl. transl. Am. Geophys. Union, Washington.]

    (b) Estimate the ratio of the area of the total leaf surface of a mature deciduous tree to the area encompassed by its root system. All other things being equal, would the transpiration from a tree be greater than from a short grass surface growing in the same area?

20. Consider the possibility of inferring an urban heat island by studying the snow cover distribution in aerial photographs. [See, "Studies on the local climate in Lund and its environs," S. Lindqvist, Dept. Geography, Royal Univ. of Lund, Sweden, *Geografiska Annaler* **50A**, 79–93 (1968).]

21. It has been suggested [H. W. Koning and Z. Jeiger, *Atm. Environ.* **2**, 615–616 (1968)] that photosynthesis decreases as ozone concentration increases. Comment on the usefulness of photosynthetic reduction as an air quality index.

22. "Studies using radio-tags are studies, not of birds, but of radio-tagged birds" (W. W. Cochran, G. G. Montgomery, and R. B. Graber, *The Living Bird* **6**, p. 224. Cornell Lab. of Ornithology, 1967). Discuss this problem, giving other examples.

23. Write an essay on sunburn. Which factor is more important, intensity of sunlight or exposure time? [See F. Daniels, Jr., J. C. Van der Leun, and B. E. Johnson, *Sci. Am.* **219**, 1, 38–46 (1969); H. E. Landsberg, "Weather and Health," Chapter 3. Doubleday, New York, 1969].

24. Consider the possibility of using lichens as an indicator of urban pollution. [See H. Schönbeck, *Staub-Reinhattung* **29**, 14–18 (1969).]

25. In Fig. 47, is it preferable to compute pollution wind roses from three anemometers or from a single reference anemometer?

26. When air temperature is above skin temperature, a human being gains heat by convection and loses heat by evaporation. Consider the effect of wind speed on the relative magnitudes of the two fluxes. Is there an optimum wind speed that maximizes cooling?

27. Figure 80 gives the ratio Q_E/Q_H plotted against temperature. Would it be preferable to use Q_H/Q_E, or does it make any difference?

28. It has been stated that a classification is required in order to understand such a complex topic as climate. Once the interrelationships are perceived, however, the classification becomes obsolete. Comment. [See F. Prohaska, *Int. J. Biometeorol.* **11**, 1–3 (1967).]

29. Undertake a library search to determine the alpine tree line around the world. Are there latitudinal or longitudinal differences? If so, can you explain them?

30. Uneven snow cover in mountainous regions has been suggested as a vegetation indicator. Conversely, vegetation patterns yield clues about the snowfall distribution. Write a short essay on this topic. [See D. N. McVean, *Weather* **13**, 197–200 (1958).]

31. Discuss the microclimate within a screened insect field cage, considering such elements as photoperiod, air temperature, leaf temperature, humidity, wind, and rainfall. Are these changes important in studies of insects? [See L. F. Hand and A. J. Keaster, *J. Econ. Entomol.* **60**, 910–915 (1967).]

32. Trees are very tall on the west coast of North America and in some parts of the humid tropics. However, there is an elfin forest in Puerto Rico at an elevation of 1050 m; the height of the canopy is only about 3–4 m in a region of abundant rainfall and much fog. Why are the trees not taller? [See H. W. Baynton, *J. Arnold Aboretum* **49**, 419–430 (1968).]

33. Is the arctic a useful environment for studying circadian rhythms? Consider separately summer conditions when the sun never sets, and the polar night. [See P. S. Corbet, *Nature* **222**, 392 (1969).]

34. Are the northern boundaries of plant, bird, animal, and insect populations affected by human settlement of the arctic and subarctic? [See M. Slessers, *Arctic* **21**, 201–203 (1968).]

35. In studies of the water balance of humans, animals, birds, and insects, should water losses be given in units of gm/gm hr or gm/cm² hr?

36. *Vapor pressure deficit* VPD is the difference between saturation vapor pressure and the actual vapor pressure at the existing air temperature. For constant absolute humidity, for example, VPD increases when the air is heated. Williams and Brochu [*Naturaliste Can.* **96**, 621–636 (1969)] have suggested that VPD is sometimes a more useful biometeorological index than absolute or relative humidity. Discuss this suggestion.

37. Develop a dimensionally correct expression for the average emission per hour of CO from automobile traffic over a short segment of highway, noting that the emission rate depends on the number of vehicles and on their speeds. Assume as a first approximation that the emission rate from an individual automobile is independent of its speed.

38. In the study of irrotational two-dimensional flows of incompressible fluids, *velocity potentials* and *stream functions* are sometimes useful. Define the terms and discuss their relevance in streamline analysis (Section 5.7).

39. M. Scharringa [*Agr. Meteorol.* **6,** 283–285 (1969)] has computed the linear regression between monthly totals of sunshine duration and monthly mean daily maximum temperature at De Bilt, Netherlands for the years 1931–1960. The correlation coefficient was $+0.81$. Yet when separate coefficients were calculated for each calendar month, values were positive for April through September, negative for December and January, and near zero for February, March, October, and November. Explain this, and discuss the usefulness of the all-data regression $(r = 0.81)$ as a predictive tool.

40. J. L. Cloudsley-Thompson [*Sci. Prog.* **56,** 499–509 (1968)] believes that the phrase "cold blooded" should not be used to describe a poikilotherm. He prefers the word "ecothermal." Discuss this point of view.

41. W. H. Terjung [*Geograph. Rev.* **60,** 31–53 (1970)] suggests that a useful biometeorological index is the environmental radiant temperature T_R averaged over sky and terrain. Explain, and describe some situations in which T_R might be quite different from "screen" air temperature.

42. J. H. Botsford (Bethlehem Steel Corp., Bethlehem, Pa.) has constructed a *wet globe thermometer,* which consists of a globe, 6.4 cm (2.5 in.) in diameter, which is covered with a moistened black cloth. The heat balance of the sphere depends on air temperature and humidity, wind speed, and the presence of radiant heat sources. Discuss biometeorological applications of this instrument.

43. Studies have been undertaken of competition in laboratory cultures of *Drosophila serrata* with *Drosophila pseudoobscura* [F. J. Ayala, *Nature* **224,** 1076–1079 (1969)]. At 25°C, *D. serrata* eliminate *D. pseudoobscura* after a few generations while at 19°C, the reverse occurs. Is there an intermediate temperature at which the two species can coexist, or must there be competitive exclusion?

44. (a) What is the physical interpretation of the ventilation coefficient (defined in Section 10.2)?
    (b) Discuss possible applications of a ventilation-coefficient wind rose.
    (c) A pollution wind rose may be constructed, similar to Fig. 45 but with radii proportional to the ventilation coefficient rather than to the wind speed. Would this representation be more meaningful than the type illustrated in Fig. 45?

45. Discuss the possible sources of errors in the use of lichens for estimat-

ing the rate of retreat of glaciers. [See, for example, J. T. Andrews and P. J. Webber, *Arctic Alpine Res.* **1**, 181–194 (1969).]

46. T. Baker [*Weather* **24**, 277–280 (1969)] has suggested that the total number of hours of sunshine in the months May to September is a simple but nevertheless meaningful summer human weather index for Britain. Do you agree?

47. O. T. Denmead [*Agr. Meteorol.* **6**, 351–371 (1969)] has compared the productivity of a wheat field and a forest. He uses the index: photosynthetic gain per unit of evaporation. Explain.

48. E. Dahl and E. Mork [*Medd. Norsk Skogförsöksvesen* **53**, 83–93 (1959)] have suggested that because the principal source of heat for growth processes (other than photosynthesis) in plants is chemical energy released by respiration, and because of the exponential nature of the relation between reaction rate and temperature [see Eq. (2.10)] when soil moisture is not limiting, a suitable growth index is formed by multiplying each hourly temperature by the appropriate respiration rate, and summing. Discuss the index and compare it with "degree days."

49. Each term in the annual heat balance and water budget of a forest can be nondimensionalized by dividing by the corresponding value in adjacent open grassland. Describe possible uses for these ratios on the local, regional and global scales. [See G. Flemming, *Wiss. Zeit. (Dresden)* **17**, 1415–1420 (1968).]

50. V. N. Adamenko and K. Sh. Khairullin [*Tr. Main Geophys. Obs.* **248**, 74–81 (1929)] have used the temperature of the exposed face as an index of wind chill. Based on experimental data obtained from groups 5–10 persons aged 19–35 years, they have obtained the following empirical results:

$$T_c = 0.4T_A - 3.3\sqrt{u} + 19$$
$$T_N = 0.4T_A - 3.3\sqrt{u} + 17$$
$$T_E = 0.4T_A - 3.3\sqrt{u} + 12$$

where T_c, T_N, and T_E are the temperatures (°C) of the cheek, nose, and ear surfaces, respectively, T_A is air temperature (°C) and u is wind speed (m/sec).

(a) If these equations are assumed to be dimensionally homogeneous, what are the dimensions of the coefficient of $\sqrt{u}$? Can you provide a physical intepretation?

(b) Using T_A as abscissa and u as ordinate, plot the lines $T_c = 0°C$, $T_N = 0°C$, $T_E = 0°C$.

(c) Compare the formulas with Eqs. (9.4) and (9.5), p. 189.

REFERENCES

1. S. N. Linzon, Locomotive smoke damage to jack pine. *Forestry Chron.* **37,** 102–106 (1961).
2. P. Buell and J. E. Dunn, Relative impact of smoking and air pollution on lung cancer. *Arch. Environ. Health* **15,** 291–297 (1967).
3. H. P. Roth, Acute smog mortality versus annual death rate cycles in certain diseases. *Extended Abstr. Intern. Biometeorol. Conf., 4th, Rutgers Univ., N. J.* (1966).
4. O. G. Edholm, Problems of acclimatization in man. *Weather* **21,** 340–350 (1966).
5. J. E. Begg, J. F. Bierhuizen, E. R. Lemon, D. K. Misra, R. O. Slatyer, and W. R. Stern, Diurnal energy and water exchanges in bulrush millet in an area of high solar radiation. *Agr. Meteorol.* **1,** 294–312 (1964).
6. D. M. Gates, "Energy Exchange in the Biosphere." Harper, New York, 1962.
7. V. P. Dadykin, On the relation between external conditions and the exchange of energy in plants of the far north. *Problemy Severa* **8,** 88–95 (English transl. from the Russian by Natl. Res. Council, Ottawa) (1964).
8. G. Harris, Climatic changes since 1860 affecting European birds. *Weather* **19,** 70–79 (1964).
9. A. J. de Villiers and J. P. Windish, Lung cancer in a fluorspar mining community. *Brit. J. Ind. Med.* **21,** 94–109 (1964).
10. R. E. Munn, "Descriptive Micrometeorology." Academic Press, New York (1966).
11. J. A. Larsen, The role of physiology and environment in the distribution of Arctic plants. Tech. Rept. No. 16, Dept. Meteorol., Univ. of Wisconsin, Madison, Wisconsin (1964).
12. L. Dounin-Barkovsky, Nonbeneficial vegetation and water resources. *Int. Assoc. Sci. Hydrol. Bull.* **10,** 48–51 (1965).
13. Biometeorology today and tomorrow. *Bull. Am. Meteorol. Soc.* **48,** 378–393 (1967).
14. J. R. Ashworth, The influence of smoke and hot gases from factory chimneys on rainfall. *Quart. J. Roy. Meteorol. Soc.* **55,** 341–350 (1929).
15. The automobile and air pollution, Part II. U. S. Dept. of Commerce, Washington, D. C. (1967).

16. F. Pasquill, "Atmospheric Diffusion." Van Nostrand, Princeton, New Jersey, 1962.
17. J. M. Mitchell, Climatic change. Tech. Note No. 79, World Meteorol. Organ., Geneva (1966).
18. F. B. Muller, J. G. Gervais, and R. W. Shaw, The effect of precipitation on the level of Lake Michigan/Huron. Rept. No. Tec-576. Meteorol. Branch, Toronto (1965).
19. Pao-K. Chang, Methods of evaluating and comparing cold waves, with special reference to New York City. *Bull. Am. Meteorol. Soc.* **30**, 107–109 (1949).
20. A. Court, Wind chill. *Bull. Am. Meteorol. Soc.* **29**, 487–493 (1948).
21. R. S. Scorer, The cost in Britain of air pollution from different types of source. Preprint Inst. of Fuel, London (1956).
22. R. I. Larsen, C. E. Zimmer, D. A. Lynn, and K. G. Blemel, Analyzing air pollutant concentration and dosage data. *J. Air Pollution Control Assoc.* **17**, 85–93 (1967).
23. E. K. Harris and S. D. Dubey, Estimating the frequency distribution of dosage from a continuous record of pollutant concentration. *Intern. J. Air Water Pollution* **8**, 369–380 (1964).
24. V. R. Evans, Electrochemical mechanism of atmospheric rusting. *Nature* **206**, 980–982 (1965).
25. I. A. Singer, An objective method for site evaluation. *J. Air Pollution Control Assoc.* **10**, 219–221 (1960).
26. R. H. Green, Estimation of tolerance over an indefinite time period. *Ecology* **46**, 887 (1965).
27. C. S. Brandt and W. W. Heck, Effects of air pollutants on vegetation. *In* "Air Pollution" (A. C. Stern, ed.), 2nd ed., Vol. 1, p. 401–443. Academic Press, New York (1968).
28. W. W. Heck, J. A. Dunning, and I. J. Hindawi, Ozone: nonlinear relation of dose and injury in plants. *Science* **151**, 577–578 (1966).
29. S. Duckworth and E. Kupchanko, Air analysis, the standard dosage–area product. *J. Air Pollution Control Assoc.* **17**, 379–383 (1967).
30. G. T. Csanady, The dosage–area problem in turbulent diffusion. *Atmos. Environ.* **1**, 451–459 (1967).
31. Report of the United Nations Scientific Committee on the Effects of Atomic Radiation. General Assembly, 21st Session, Suppl. No. 14 (A/6314) (1966).
32. Report of the United Nations Scientific Committee on the Effects of Atomic Radiation. General Assembly, 17th Session, Suppl. No. 16 (A/5216) (1962).
33. R. D'Have and H. Motteu, The necessity of cooperation between meteorologists and the construction industry: Belgian experience. *Preprint Symp. World Meteorol. Organ. Urban Climate Building Climatology* (1968).
34. N. Marshall, The icefields round Iceland in spring 1968. *Weather* **23**, 368–376 (1968).
35. W. H. Hogg, Meteorology and horticulture. *Weather* **19**, 234–241 (1964).
36. S. Brody, Climatic physiology of cattle. *J. Dairy Sci.* **39**, 715–725 (1956).
37. F. F. Davitaya, A method of predicting heat supply and duration of the growth period. *Agr. Meteorol.* **2**, 109–119 (1965).
38. M. Y. Nuttonson, The role of bioclimatology in agriculture with special reference to the use of thermal and photothermal requirements of pure-line varieties of plants as a biological indicator in ascertaining climatic analogues. *Int. J. Bioclimatol. Biometeorol.* **1** (Part II, Sec. B 2), 1–20 (1957).

39. R. Lee and W. E. Sharpe, Time-integrated thermal effects of forest irrigation. *Preprint Conf. Agr. Meteorol., 8th, Ottawa, Ont.*, Am. Meteorol. Soc. (1968).
40. V. Yevjevich, An objective approach to definitions and investigations of continental hydrologic droughts. Hydrology Papers No. 23, Colorado State Univ. (1967).
41. F. D. H. Macdowall, E. I. Mukammal, and A. F. W. Cole, Direct correlation of air-polluting ozone and tobacco weather fleck. *Can. J. Plant Sci.* **44**, 410–417 (1964).
42. E. I. Mukammal, Ozone as a cause of tobacco injury. *Agr. Meteorol.* **2**, 145–165 (1965).
43. A. S. Gurvich and T. K. Kravchenko, The frequency spectrum of small-scale temperature fluctuations. *Bull. Acad. Sci. USSR Inst. At. Phys.* **4**, 144–146 (English transl. by Am. Meteorol. Soc. AF 19(628)-3880) (1962).
44. G. C. Gill, On the dynamic response of meteorological sensors and recorders. *Proc. Can. Conf. Micrometeorol., 1st* pp. 1–27. Meteorol. Branch, Toronto (1967).
45. H. A. Panofsky and G. W. Brier, "Some Applications of Statistics to Meteorology." College of Mineral Industries, Pennsylvania State Univ., University Park, Pennsylvania (1958).
46. R. I. Larsen, F. B. Benson, and G. A. Jutze, Improving the dynamic response of continuous air pollutant measurements with a computer. *J. Air Pollution Control Assoc.* **15**, 19–22 (1965).
47. T. Takakura, Predicting air temperatures in the glasshouse. *J. Meteorol. Soc. Japan* **45**, 40–52 (1967).
48. D. Bryant, An investigation into the response of thermometer screens—the effect of wind speed on the lag time. *Meteorol. Mag.* **97**, 183–186 (1968).
49. R. Crawford and H. S. Ward, Determination of the natural periods of buildings. *Bull. Seismol. Soc.* **54**, 1743–1756 (1964).
50. J. Aschoff, Adaptive cycles: their significance for defining environmental hazards. *Intern. J. Bioclimatol. Biometeorol.* **11**, 255–278 (1967).
51. G. E. Fogg, "The Growth of Plants." Penguin Books, Harmondsworth, England, 1963.
52. C. C. Davis, Circadian, and related rhythms. *Ecology* **48**, 171 (1967).
53. L. Curry, Climatic change as a random series. *Ann. Assoc. Am. Geographers* **52**, 21–31 (1962).
54. L. T. Evans, Extrapolation from controlled environments to the field. *In* "Environmental Control of Plant Growth" (L. T. Evans, ed.), pp. 421–435. Academic Press, New York (1963).
55. N. Stark, Spring transpiration of three desert species. *J. Hydrol.* **6**, 297–305 (1968).
56. J. H. Fremlin, Nature's time-scale. *Nature* **211**, 1107–1108 (1966).
57. D. F. Parkhurst and D. M. Gates, Transpiration resistance and energy budget of *Populus sargentii* leaves. *Nature* **210**, 172–174 (1966).
58. H. Ryd, The importance of meteorology in building. Preprint CLU/Doc. 30, *Symp. Urban Climates Building Climatol. Brussels,* World Meteorol. Organ., Geneva (1968).
59. Sir O. G. Sutton, Micrometeorology. *Sci. Am.* **211**, 62–76 (1964).
60. D. M. Gates, Energy exchange in the biosphere. *Proc. Symp. Functioning of Terrestrial Ecosystems at the Primary Production Level* pp. 33–43. UNESCO, Paris (1968).

61. E. I. Mukammal and C. J. Baker, Effective radiant energy in a red pine forest. *Preprint Conf. Agr. Meteorol., 8th, Ottawa, Ont.,* Am. Meteorol. Soc. (1968).
62. K. P. Chopra and L. F. Hubert, Karman vortex-streets in the earth's atmosphere. *Nature* **203,** 1341–1343 (1964).
63. J. N. Hunt and P. E. Wickins, Vortex streets in the earth's atmosphere. *Geofis. Pura. Appl. Geophys.* **67,** 179–185 (1967).
64. W. Brinkmann and I. Y. Ashwell, The structure and movement of the Chinook in Alberta. *Atmosphere* **6** (2), 1–10 (1968).
65. R. E. Munn and D. Storr, Meteorological studies in the Marmot Creek watershed, Alberta, Canada in August 1965. *Water Resources Res.* **3,** 713–722 (1967).
66. D. B. Smith, Tracer study in an urban valley. *J. Air Pollution Control Assoc.* **18,** 466–471 (1968).
67. H. E. Landsberg, Critique of certain climatological procedures. *Bull. Am. Meteorol. Soc.* **28,** 187–191 (1947).
68. J. B. Wright, Precipitation patterns over Vancouver city and lower Fraser valley. Rept. No. Tec-623. Meteorol. Branch, Toronto (1966).
69. F. K. Hare, Some climatological problems of the arctic and sub-arctic. *In* "Compendium of Meteorology," pp. 952–964. Am. Meteorol. Soc., Boston, Massachusetts, 1951.
70. R. A. Bryson, Airmasses, streamlines, and the boreal forest. Tech. Rept. No. 24. Dept. of Meteorol., Univ. of Wisconsin, Madison, Wisconsin (1966).
71. A. M. Obukhov, Turbulence and weather. *In Proc. IAMAP Gen. Assembly, 14th, Lucerne* pp. 102–107. Intern. Assoc. Meteorol. At. Phys. (available from Dr. W. L. Godson, Meteorol. Branch, Toronto) (1968).
72. H. E. Landsberg, Meteorological observations in urban areas. *Preprint Symp. Meteorol. Observations Instrumentation,* Ob. 2.2. Am. Meteorol. Soc., Boston, Massachusetts (1969).
73. Rules and Regulations, Title 10, Code of Federal Regulations, U. S. At. Energy Comm., Government Printing Office, Washington, D. C. (1965).
74. D. M. Keagy, W. W. Stalker, C. E. Zimmer, and R. C. Dickerson, Sampling station and time requirements for urban air pollution surveys, Part I, *J. Air Pollution Control Assoc.* **11,** 270–280 (1961).
75. W. W. Stalker and R. C. Dickerson, Sampling station and time requirements for urban air pollution surveys. Part III. *J. Air Pollution Control Assoc.* **12,** 170–178 (1962).
76. W. W. Stalker, R. C. Dickerson, and G. D. Kramer, Sampling station and time requirements for urban air pollution surveys. Part IV. *J. Air Pollution Control Assoc.* **12,** 361–375 (1962).
77. M. Clifton, D. Kerridge, W. Moulds, J. Pemberton, and J. K. Donoghue, The reliability of air pollution measurements in relation to the siting of instruments. *Intern. J. Air Pollution* **2,** 188–196 (1959).
78. A. E. Carte, Areal hail frequency. *J. Appl. Meteorol.* **6,** 336–338 (1967).
79. J. Idrac, Caractéristiques d'une chaîne de mesure. *Proc. Symp. Functioning of Terrestrial Ecosystems at the Primary Production Level* pp. 151–175. UNESCO, Paris (1968).
80. K. D. Hage, C. H. H. Diehl, and M. G. Dudley, On horizontal flat-plate sampling of solid particles in the atmosphere. *AMA Arch. Ind. Health* **21,** 124–131 (1960).
81. J. B. Harrington, G. C. Gill, and B. R. Warr, High-efficiency pollen samplers for use in clinical allergy. *J. Allergy* **30,** 357–375 (1959).

82. P. A. Leighton, W. A. Perkins, S. W. Grinnell, and F. X. Webster, The fluorescent particle atmospheric tracer. *J. Appl. Meteorol.* **4**, 334–348 (1965).
83. A. N. Dingle, Hay fever pollen counts and some weather effects. *Bull. Am. Meteorol. Soc.* **38**, 465–469 (1957).
84. G. Stanhill, Rainfall measurements at ground level. *Weather* **13**, 33–34 (1958).
85. J. Glasspoole, Rainfall measurements at ground level. *Weather* **13**, 211 (1958).
86. "Handbook of Meteorological Instruments." H. M. Stationery Office, London, 1956.
87. R. E. Lacy, Distribution of rainfall round a house. *Meteorol. Mag.* **80**, 184–189 (1951).
88. A. L. Cole and J. P. Harrington, Atmospheric dispersion of ragweed pollen. *J. Air Pollution Control Assoc.* **17**, 654–656 (1967).
89. W. P. Pruitt, F. Lourence, and T. V. Crawford, Radiation and energy balance changes during the eclipse of 20 July 1963. *J. Appl. Meteorol.* **4**, 272–278 (1965).
90. R. S. Scorer, The nature of convection as revealed by soaring birds and dragon-flies. *Quart. J. Roy. Meteorol. Soc.* **80**, 68–77 (1954).
91. T. Okita, Estimation of direction of air flow from observation of rime ice. *J. Met. Soc. Japan* **38**, 207–209 (1960).
92. T. Okita, Some chemical and meteorological measurements of air pollution in Asahikawa. *Intern. J. Air Water Pollution* **9**, 323–332 (1965).
93. J. Jenik and J. B. Hall, The ecological effects of the harmattan wind in the Djebobo Massif (Togo mountains, Ghana). *J. Ecology* **54**, 767–779 (1966).
94. T. Sekiguti, Prevailing winds in early summer and bending shapes of persimmon trees at Akaho Fan, Nagano Prefecture, Japan. *Tokyo J. Climatol.* **2**, 13–25 (1965).
95. D. Thomas, The assessment of shelter-need upon exposed farm sites. *Weather* **14**, 375–384 (1959).
96. D. Thomas, The tattering rate of flags as an index of exposure to wind. *Meteorol. Mag.* **88**, 67–70 (1959).
97. N. Rutter, Tattering of flags at different sites in relation to wind and weather. *Agr. Meteorol.* **5**, 163–181. (1968).
98. T. Shibano, Local weather in southern Boso peninsula. Estimation of prevailing winds on the basis of landscape. *J. Meteorol. Res. (Japan)* **18**, 329–337 (in Japanese) (1966).
99. J. N. Myers, The use of vegetation in the control of shallow radiation fog. *Weather* **22**, 289–291 (1967).
100. W. W. Heck, The use of plants as indicators of air pollution. *Intern. J. Air Water Pollution* **10**, 99–111 (1966).
101. H. C. McKee and F. W. Bieberdorf, Vegetation symptoms as a measure of air pollution. *J. Air Pollution Control Assoc.* **10**, 222–225 (1960).
102. E. P. Jeffree, Some long-term means from the Phenological Reports (1891–1948) of the Royal Meteorological Society. *Quart. J. Roy. Meteorol. Soc.* **86**, 95–103 (1960).
103. H. Arakawa, Climatic change as revealed by the freezing dates of Lake Suwa in central Japan. *J. Meteorol.* **12**, 94 (1955).
104. H. Arakawa, Climatic change as revealed by the blooming dates of the cherry blossoms at Kyoto. *J. Meteorol.* **13**, 599–600 (1956).
105. R. Geiger, "The Climate Near the Ground." Harvard Univ. Press, Cambridge, Massachusetts. (English transl. from the German by Harvard Univ. Press) (1965).

106. M. T. Jackson, Effects of microclimate on spring flowering phenology. *Ecology* **47**, 407–415 (1966).

107. M. Kalb, Einige Beiträge zum Stadtklima von Köln. *Meteorolog. Runds.* **15**, 92–99 (1962).

108. A. A. Levashov, Approximate determination of high flood frequency in rivers without hydrological observations. *Meteorol. i Gidrol.* **10**, 48–49 (Engl. transl. by Am. Geophys. Union, Soviet Hydrology: selected papers, No. 5, 1966) (1966).

109. V. A. Saull, Examples of natural photography from Renfrew County, Ont. *Can. J. Earth Sci.* **4**, 619–623 (1967).

110. J. Sherrod and H. Neuberger, Understanding forecast terms—results of a survey. *Bull. Am. Meteorol. Soc.* **39**, 34–36 (1958).

111. J. J. Phair, G. C. R. Carey, R. J. Shephard, and M. L. Thomson, Some factors in the design, organization and implementation of an air hygiene survey. *Intern. J. Air. Pollution* **1**, 18–30 (1958).

112. D. O. Anderson, B. G. Ferris, and T. W. Davis, The Chilliwack respiratory survey, 1963. *Can. Med. Assoc. J.* **92**, 899–905 (1965).

113. J. Schusky, L. Goldner, S. Z. Mann, and W. C. Loring, Methodology for the study of public attitudes concerning air pollution. *J. Air Pollution Control Assoc.* **14**, 445–448 (1964).

114. D. O. Anderson and B. G. Ferris, Community studies of the health effects of air pollution—a critique. *J. Air Pollution Control Assoc.* **15**, 587–593 (1965).

115. W. S. Smith, J. J. Schueneman, and L. D. Zeidberg, Public reaction to air pollution in Nashville, Tennessee. *J. Air Pollution Control Assoc.* **14**, 418–423 (1964).

116. H. J. Paulus and T. J. Smith, Association of allergic bronchial asthma with certain air pollutants and weather parameters. *Intern. J. Biometeorol.* **11**, 119–127 (1967).

117. J. McCarroll, Measurements of morbidity and mortality related to air pollution. *J. Air Pollution Control Assoc.* **17**, 203–209 (1967).

118. C. A. Bridger and L. A. Helfland, Mortality from heat during July 1966 in Illinois. *Intern. J. Biometeorol.* **12**, 51–69 (1968).

119. K. A. Bouchtoueva, Methods for studies of the effect of atmospheric pollution on the population. *Gigiena i Sanit.* **6**, 93–98 (1966).

120. J. Akerman, Indoor climate. Seminar on human biometeorology, pp. 133–144, U. S. Dept. of Health, Education and Welfare, Publ. Health Service Publ. 999-AP-25, Washington, D. C. (1967).

121. I. Andersen, Medical-hygienic evaluation of indoor climate. *Preprint Symp. World Meteorol. Organ. Urban Climate and Building Climatol. Geneva* (1968).

122. R. L. Desjardins and G. W. Robertson, Variations of meteorological factors in a greenhouse. *Can. Agr. Eng.* **10**, 85–90 (1968).

123. L. A. Hunt, I. I. Impens, and E. R. Lemon, Preliminary wind tunnel studies of the photosynthesis and evapotranspiration of forage stands. *Crop Sci.* **7**, 575–578 (1967).

124. R. N. Morse and L. T. Evans, Design and development of CERES—an Australian phytotron. *J. Ag. Engin. Res.* **7**, 128–140 (1962).

125. D. F. Adams, An air pollution phytotron. *J. Air Pollution Control Assoc.* **11**, 470–476 (1961).

126. H. A. Senn, D. P. Anderson, and L. C. Anderson, Biotron manual for investigators. Univ. of Wisconsin, Madison, Wisconsin (1965).

127. D. E. Reichle, The temperature and humidity relations of some bog pselaphid beetles. *Ecology* **48**, 208–215 (1967).
128. H. Buchberg, K. W. Wilson, M. H. Jones, and K. G. Lindh, Studies of interacting atmospheric variables and eye irritation thresholds. *Intern. J. Air Water Pollution* **7**, 257–280 (1963).
129. J. L. Hollander and S. J. Yeostros, The effect of simultaneous variations of humidity and barometric pressure on arthritis. *Bull. Am. Meteorol. Soc.* **44**, 489–494 (1963).
130. V. Olgyay, "Design with Climate." Princeton Univ. Press, Princeton, New Jersey (1963).
131. L. K. Paulsell and D. B. Lawrence, Artificial frost apparatus. *Ecology* **44**, 146–148 (1963).
132. F. Spierings, Method for determining the susceptibility of trees to air pollution by artificial fumigation. *Atmos. Environ.* **1**, 205–210 (1967).
133. D. H. K. Lee and J. A. Vaughan, Temperature equivalent of solar radiation on man. *Intern. J. Biometeorol.* **8**, 61–69 (1964).
134. W. Larcher, Physiological approaches to the measurement of photosynthesis in relation to dry matter production of trees. *Preprint Conf. Nat. Agrometeorol., 8th,* Am. Meteorol. Soc., Boston, Massachusetts (1968).
135. A. D. McFadden, Effect of seed source on comparative test results in barley. *Can. J. Plant Sci.* **43**, 295–300 (1963).
136. J. P. Grime, Comparative experiments as a key to the ecology of flowering plants. *Ecology* **46**, 513–515 (1965).
137. J. R. Clements, Distribution of rainfall and soil moisture depletion in a forest clearing. *Bull. Am. Meteorol. Soc.* **47**, 465 (1966).
138. P. J. B. Duffy and J. W. Fraser, Local frost occurrences in eastern Ontario woodlands. Publ. No. 1029, Forest Res. Branch, Dept. of Forestry, Ottawa (1963).
139. E. L. Hawke, Frost hollows. *Weather* **1**, 41–45 (1946).
140. J. R. Clements, Solar radiation in a forest clearing. *Weather* **21**, 316–317 (1966).
141. J. E. Carson, R. Votruba, and J. Lin, Chicago fuel-switch test. *Preprint Annual Meeting, Air Pollution Control Assoc., 62nd, Pittsburgh, Pa.* (1969).
142. D. F. Gatz, A. N. Dingle, and J. W. Winchester, Detection of indium as an atmospheric tracer by neutron activation. *J. Appl. Meteorol.* **8**, 229–235 (1969).
143. D. A. Fraser and E. E. Gaertner, Utilization of radiosotopes in forestry research, *Preprint Congr. World Forestry, 6th, Madrid* (1966).
144. G. W. Robertson, A biometeorological time scale for a cereal crop involving day and night temperatures and photoperiod. *Intern. J. Biometeorol.* **12**, 191–223 (1968).
145. I. A. Singer and G. S. Raynor, A solar time classification for meteorological use. *Bull. Am. Meteorol. Soc.* **39**, 569–573 (1958).
146. C. E. P. Brooks and N. Carruthers, "Handbook of Statistical Methods in Meteorology." H. M. Stationery Office, London, 1953.
147. R. E. Munn and C. R. Ross, Analysis of smoke observations at Ottawa, Canada. *J. Air Pollution Control Assoc.* **11**, 410–416 (1961).
148. A. F. Chisholm and F. B. Muller, Mesometeorology and short range forecasting. Rept. No. Tec-620. Meteorol. Branch, Toronto (1966).
149. J. C. Bellamy, Data display for analysis. Symp. Environmental Measurements pp. 213–226. Public Health Service Publ. No. 999-AP-15, Cincinnati, Ohio (1964).

150. J. I. P. Jones, Presentations of surface wind and turbulence using the cathode-ray tube. *J. Appl. Meteorol.* **5**, 25–32 (1966).
151. J. H. Emslie, Wind flow in Burrard Inlet, Vancouver, B. C. Rept. No. Tec-686. Meteorol. Branch, Toronto (1968).
152. Air quality criteria for sulfur oxides. U. S. Dept. of Health, Education, and Welfare, Public Health Service, Washington, D. C. (1967).
153. J. M. Hirst, O. J. Stedman, and H. W. Hogg, Long-distance spore transport: methods of measurement, vertical spore profiles and the detection of immigrant spores. *J. Gen. Microbiol.* **48**, 329–355 (1967).
154. J. R. Wallis, Multivariate statistical methods in hydrology—comparison using data of known functional relationship. *Water Resources Res.* **1**, 447–461 (1965).
155. S. P. Jackson, Climatological atlas of Africa. Commission for Tech. Co-operation in Africa. Joint Climatology Unit, Univ. of Witwatersrand, Johannesburg, South Africa (55 plates) (1961).
156. F. Steinhauser, Methods of evaluation and drawing of climatic maps in mountainous countries. *Arch. Meteorol. Geophys. Bioklimatol.* **15B**, 329–358 (1967).
157. J. P. Kerr, G. W. Thurtell, and C. B. Tanner, Mesoscale sampling of global radiation analysis of data from Wisconsin. *Monthly Weather Rev.* **96**, 237–241 (1968).
158. Guide to Climatological Practices. WMO-No. 100. TP. 44, World Meteorol. Organ., Geneva (1960).
159. R. E. Munn, D. A. Thomas, and A. F. W. Cole, A study of suspended particulate and iron concentrations in Windsor, Canada. *Atmos. Environ.* **3**, 1–10 (1969).
160. J. Glasspoole, Assessment of areal rainfall amounts. *Weather* **17**, 312–313 (1962).
161. A. H. Thiessen, Precipitation averages for large areas. *Monthly Weather Rev.* **39**, 1082–1084 (1911).
162. P. Haggett, Towards a statistical definition of ecological range: the case of *Quercus suber*. *Ecology* **45**, 622–625 (1964).
163. W. Böer, "Technische Meteorologie." Teubner, Leipzig, 1964.
164. R. Reidat, Climatological data for rain penetration. *Preprint Symp. World Meteorol. Organ. Urban Climate Building Climatol.* (1968).
165. G. C. Holzworth, A note on surface wind-speed observations. *Monthly Weather Rev.* **93**, 323–325 (1965).
166. L. Truppi, Bias introduced by anemometer starting speeds in climatological wind rose summaries. *Monthly Weather Rev.* **96**, 325–327 (1968).
167. E. W. Hewson, G. C. Gill, and H. W. Baynton, Meteorological analysis. Progr. Rept., 3rd, UMRI 2515-3-P. Dept. of Civil Eng., Univ. of Michigan, Ann Arbor, Michigan (1959).
168. C. E. Wallington, A method of reducing observing and procedure bias in wind-direction frequencies. *Meteorol. Mag.* **97**, 293–302 (1968).
169. A. Court, Wind roses. *Weather* **18**, 106–110 (1963).
170. J. McCormick and P. A. Harcombe, Phytograph: useful tool or decorative doodle? *Ecology* **49**, 13–20 (1968).
171. R. V. Dexter, The sea-breeze hodograph at Halifax. *Bull. Am. Meteorol. Soc.* **39**, 241–247 (1958).
172. H. L. Crutcher, On the standard vector-deviation wind rose. *J. Meteorol.* **14**, 28–33 (1957).

173. N. Untersteiner, Eine neue Methode der Darstellung Kleinräumiger Windfelder für praktisch-klimatologische Zwecke. *Arch. Meteorol. Geophys. Bioklimatol.* **B10**, 222–227 (1960).
174. K. Cehak, Some examples of climatological data processing for technical purposes. *Preprint Symp. World Meteorol. Organ. Urban Climate Building Climatol.* (1968).
175. Parkersburg, West Virginia–Marietta, Ohio air pollution abatement activities. Tech. Rept., Natl. Center for Air Pollution Control, U. S. Dept. Health, Education, and Welfare, Cincinnati, Ohio (1967).
176. S. Petterssen, "Weather Analysis and Forecasting," Vol. I. McGraw-Hill, New York, 1956.
177. H. B. Schultz and M. Fitzwater, Influence of "seabreeze" on the wind pattern in an agricultural valley 100 miles inland. Paper presented at *Natl. Conf. Agr. Meteorol., 8th,* Am. Meteorol. Soc., Boston, Massachusetts (1968).
178. J. M. Hirst and G. W. Hurst, Long-distance spore transport. Airborne Microbes, *(Symp. Soc. Gen. Microbiol.),* No. XVII, pp. 307–344 (1967).
179. C. S. Durst and N. E. Davis, Accuracy of geostrophic trajectories. *Meteorol. Mag.* **86**, 138–141 (1957).
180. H. P. Sanderson, P. Bradt, and M. Katz, A study of dustfall on the basis of replicated Latin Square arrangements of various types of collectors. *J. Air Pollution Control Assoc.* **13**, 461–466 (1963).
181. A. H. Robinson and R. A. Bryson, A method for describing quantitatively the correspondence of geographical distributions. *Ann. Assoc. Am. Geograph.* **47**, 379–391 (1957).
182. H. A. Panofsky, Significance of meteorological correlation coefficients. *Bull. Am. Meteorol. Soc.* **30**, 326–327 (1949).
183. V. M. Yevjevich, Discussion. *Proc. Symp. Hydrol., 5th,* p. 34. Natl. Res. Council, Ottawa (1966).
184. J. Schubert, A. Brodsky, and S. Tyler, The log-normal function as a stochastic model of the distribution of strontium-90 and other fission products in humans. *Health Phys.* **13**, 1187–1204 (1967).
185. C. E. Zimmer and R. I. Larsen, Calculating air quality and its control. *J. Air Pollution Control Assoc.* **15**, 565–572 (1965).
186. H. Stratmann and D. Rosin, Untersuchungen über die Bedeutung einer empirischen Kenngrösse zur Beschreibung der Häufigkeitsverteilung von SO₂-Konzentrationen in der Atmosphäre. *Staub-Reinhalt* **24**, 520–525 (1964).
187. J. Juda, Planung und Auswertung von Messungen der Verunreinigungen in der Luft. *Staub-Reinhalt,* **28**, 186–192 (1968).
188. G. R. Lundqvist, Ideal response graphs for psychophysical judgments in environmental hygiene with special reference to odor scaling. Symp. Theories of Odors and Odor Measurement (N. Tanyolac, ed.), pp. 447–450. Robert College Res. Center, Bebek, Istanbul, Turkey (1968).
189. D. V. Anderson, Review of basic statistical concepts in hydrology. *Proc. Symp. Hydrol., 5th* pp. 3–27. Natl. Res. Council, Ottawa (1966).
190. J. V. Nou, Development of the diffusion prediction equation. The Ocean Breeze and Dry Gulch Diffusion Programs (D. A. Haugen and J. H. Taylor, eds.), Vol. II, pp. 1–21, AFCRL-63-791. Air Force Cambridge Res. Labs., Massachusetts (1963).
191. C. K. Stidd, The use of eigenvectors for climatic estimates. *J. Appl. Meteorol.* **6**, 255–264 (1967).

192. W. D. Sellers, Climatology of monthly precipitation patterns in the western United States, 1931–1966. *Monthly Weather Rev.* **96,** 585–595 (1968).

193. M. Grimmer, The space-filtering of monthly surface anomaly data in terms of pattern, using empirical orthogonal functions. *Quart. J. Roy. Meteorol. Soc.* **89,** 395–408 (1963).

194. N. C. Matalas and B. J. Reiher, Some comments on the use of factor analysis. *Water Resources Res.* **3,** 213–223 (1967).

195. G. R. Kendall, Statistical analysis of extreme values. *Proc. Symp. Hydrol., 1st* pp. 54–83. Natl. Res. Council, Ottawa (1959).

196. E. J. Gumbel, "Statistics of Extremes." Columbia Univ. Press, New York, 1958.

197. I. I. Gringorten, Fitting meteorological extremes by various distributions. *Quart. J. Roy. Meteorol. Soc.* **88,** 170–176 (1962).

198. E. J. Gumbel, Extreme value analysis of hydrologic data. *Proc. Symp. Hydrol., 5th* pp. 147–181. Natl. Res. Council, Ottawa (1966).

199. Z. Kaczmarek, Efficiency of the estimation of floods with a given return period, Intern. Assoc. Sci. Hydrol. Publ. No. 45, 3, 144–159. General Assembly, Toronto (1957).

200. H. C. S. Thom, Toward a universal climatological extreme wind distribution. *In* "Wind Effects on Buildings and Structures," Vol. 1, pp. 669–683. Univ. of Toronto Press, Toronto, 1968.

201. G. A. Tunnell, Discussion. *J. Roy Statist. Soc.* **A120,** 428–430 (1957).

202. M. A. Benson, Average probability of extreme events. *Water Resources Res.* **3,** 225 (1967).

203. I. A. Singer, The relationship between peak and mean concentrations. *J. Air Pollution Control Assoc.* **11,** 336–341 (1961).

204. F. A. Gifford, Statistical properties of a fluctuating plume dispersion model. *Advan. Geophys.* **6,** 117–138 (1959).

205. F. A. Gifford, Peak to average concentration ratios according to a fluctuating plume dispersion model. *Intern. J. Air Water Pollution* **3,** 253–260 (1960).

206. I. A. Singer, K. Imai, and R. G. Del Campo, Peak to mean pollutant concentration ratios for various terrain and vegetation cover. *J. Air Pollution Control Assoc.* **13,** 40–42 (1963).

207. W. T. Hinds, On the variance of concentration in plumes and wakes. BNWL-SA-1435, Battelle Memorial Inst., Pacific Northwest Lab., Richland, Washington (1967).

208. Y. Mitsuta, Gust factor and analysis time of gust. *J. Met. Soc. Japan* **40,** 242–244 (1962).

209. G. T. Csanady, Concentration fluctuations in turbulent diffusion. *J. At. Sci.* **24,** 21–28 (1967).

210. R. B. Faoro, A study of sample averaging times and peak-to-mean ratios for gaseous pollutants. Preprint Annual Meeting Air Pollution Control Assoc., Toronto (1965).

211. P. J. Barry, The concept of a standard site. AECL-2682, At. Energy of Canada Ltd., Chalk River, Ontario (1967).

212. H. H. Lamb, Our changing climate, past and present. *Weather* **14,** 299–318 (1959).

213. J. M. Craddock and M. Grimmer, The estimation of mean annual temperature from the temperatures of preceding years. *Weather* **15,** 340–348 (1960).

214. A. Court, Climatic normals as predictors, Sci. Rept. No. 1, AF 19(628)–5716, Air Force Cambridge Res. Labs., Bedford, Massachusetts (1967).

215. H. E. Landsberg and W. C. Jacobs, Applied climatology. Compendium of Meteorol. pp. 976–992. Am. Meteorol. Soc., Boston, Massachusetts (1951).

216. H. C. S. Thom, Some methods of climatological analysis. Tech. Note No. 81. World Meteorol. Organ., Geneva (1966).

217. N. C. Matalas, Some aspects of time series analysis in hydrologic studies. *Proc. Symp. Hydrol. 5th* pp. 271–309. Natl. Res. Council, Ottawa (1966).

218. R. B. Blackman and J. W. Tukey, "The Measurement of Power Spectra." Dover, New York (1959).

219. F. B. Muller, Mesometeorology and short-range forecasting—Rept. No. 1. Can. Meteorol. Memoir No. 24, Meteorol. Branch, Toronto (1966).

220. K. D. Sabinin, Selection of the relation between periodicity of measurement and instrument inertia in sampling. *Izv. Atm. Ocean. Phys.* **3**, 973–980 [English transl. by Am. Geophys. Union, **3**, 268–272 (1967)].

221. J. C. Kaimal, The effect of vertical line averaging of the spectra of temperature and heat flux. *Quart. J. Roy. Meteorol. Soc.* **94**, 149–155 (1968).

222. V. G. Alekseev, The statistical properties of spectrum analyzer measurements. *Izv. Atm. Ocean. Phys.* **3**, 928–935 [English transl. by Am. Geophys. Union, **3**, 541–544 (1967)].

223. J. W. Cooley and J. W. Tukey, An algorithm for the machine calculation of complex Fourier series. *Math. Comput.* **19**, 297–301 (1965).

224. C. Bingham, M. D. Godfrey, and J. W. Tukey, Modern techniques of power spectrum estimation. *IEEE Trans. Audio and Electroacoustics* **15**, 56–65 (1967).

225. P. R. Julian, Variance spectrum analysis. *Water Resources Res.* **3**, 831–845 (1967).

226. J. Namias, Long range weather forecasting—history, current status and outlook. *Bull. Am. Meteorol. Soc.* **49**, 438–470 (1968).

227. J. W. Tukey, Data analysis and the frontiers of geophysics. *Science* **148**, 1283–1289 (1965).

228. J. W. A. Brant and S. R. G. Hill, Human respiratory diseases and atmospheric air pollution in Los Angeles, California. *Intern. J. Air Water Pollution* **8**, 259–277 (1964).

229. D. L. Gilman, F. J. Fuglister, and J. R. Mitchell, On the power spectrum of "red noise." *J. Atmos. Sci.* **20**, 182–184 (1963).

230. K. R. Gabriel and J. Neumann, A Markov chain model for daily rainfall occurrence at Tel Aviv. *Quart. J. Roy. Meteorol. Soc.* **88**, 90–95 (1962).

231. W. P. Lowry and D. Guthrie, Markov chains of order greater than one. *Monthly Weather Rev.* **96**, 798–801 (1968).

232. I. I. Gringorten, A stochastic model of the frequency and duration of weather events. *J. Appl. Meteorol.* **5**, 606–624 (1966).

233. I. I. Gringorten, Probabilities of moving time averages of a meteorological variate. *Tellus* **20**, 461–471 (1968).

234. D. Sharon, On the further development of Gringorten's stochastic model for climatological predictions. *J. Appl. Meteorol.* **6**, 625–630 (1967).

235. C. L. Godske, Contribution to statistical meteorology. *Geophys. Norvegica* **24**, 161–210 (Bjerknes Memorial Volume) (1962).

236. E. Eriksson, An exercise in stochastic hydrology. *Nordic Hydrol.* **1**, (1970) (in press).

237. J. Aitchison and J. A. C. Brown, "The Lognormal Distribution." Cambridge Univ. Press, London and New York, 1966.

238. W. S. Wayne, P. E. Wehrle, and R. E. Carroll, Oxidant air pollution and athletic performance. *J. Am. Med. Assoc.* **199**, 901–904 (1967).

239. E. T. Linacre, Calculations of the transpiration rate and temperature of a leaf. *Arch. Meteorol. Geophys. Bioklimatol.* **B13**, 391–399 (1965).

240. J. L. Monteith, Light distribution and photosynthesis in field crops. *Ann. Botany N.S.* **29**, 17–37 (1965).

241. H. L. Langharr, "Dimensional Analysis and Theory of Models." Wiley, New York (1951).

242. G. R. Lord and H. J. Leutheusser, Wind-tunnel modelling of stack gas discharge. *Preprint Banff Conf. Pollution,* Eng. Inst. of Canada (1968).

243. S. I. Kharchenko and K. I. Kharchenko, Total evaporation from the soil in a zone of deficient moisture and a method of its calculation. *Trans. State Hydrol. Inst.* (*Tr. GGI*) **125**, 34–57; (English transl. by Am. Geophys. Union, **125**, 511–529) (1965).

244. W. R. Dawson, V. H. Shoemaker, and P. Licht, Evaporative water losses of some small Australian lizards. *Ecology* **47**, 589–594 (1966).

245. L. L. Getz, Notes on the water balance of the redback vole. *Ecology* **43**, 565–566 (1962).

246. J. L. Cloudsley-Thompson and E. R. M. Mohamed, Water economy of the ostrich. *Nature* **216**, 1040 (1967).

247. R. E. MacMillen and A. K. Lee, Australian desert mice: independence of exogenous water. *Science* **158**, 383–385 (1967).

248. E. Lemon, Aerodynamic studies of CO_2 exchange between atmosphere and the plant. *In* "Harvesting the Sun: Photosynthesis in Plant Life" (A. San Pietro, F. A. Greer, and T. J. Army, eds., pp. 263–290. Academic Press, New York, 1967.

249. W. Koch, Des Tagesgang der Produktivität de Transpiration. *Planta* **48**, 418–452 (1957).

250. J. D. Hodges, Patterns of photosynthesis under natural environmental conditions. *Ecology* **48**, 234–242 (1967).

251. D. C. Spanner, The Peltier effect and its use in the measurement of suction pressure. *J. Exp. Botany* **2**, 145–168 (1951).

252. T. Totsuka, N. Nomoto, T. Oikawa, Y. Ino, T. Saeki, and M. Monsi, Comparison of energy balance method and half-leaf method for photosynthesis measurement in a plant community. *Preprint Conf. Agr. Meteorol., 8th* Am. Meteorol. Soc., Boston (1968).

253. R. O. Slatyer and P. G. Jarvis, Gaseous diffusion porometer for continuous measurement of diffusive resistance of leaves. *Science* **151**, 574–576 (1966).

254. W. D. P. Stewart, Nitrogen-fixing plants. *Science* **158**, 1426–1432 (1967).

255. A. J. Woodall, The physics of human comfort. *Bull. Inst. Phys.* **14**, 141–147 (1963).

256. B. K. McNab, A model of the energy budget of a wild mouse. *Ecology* **44**, 521–532 (1964).

257. J. M. Miranda, V. J. Konopinski, and R. I. Larsen, Carbon monoxide control in a high highway tunnel. *Arch. Environ. Health* **15**, 16–25 (1967).

258. H. E. Landsberg, "Weather and Health," Doubleday, New York, 1969.

259. L. R. Orlenko, Wind and its technical aspects. *Preprint Symp. Urban Climates Building Climatol.* World Meteorol. Organ., Geneva (1968).

260. P. E. Waggoner, Moisture loss through the boundary layer. *Intern. J. Biometeorol.* **3**, 41–52 (1967).

261. I. R. Cowan, Transport of water in the soil–plant–atmosphere system. *J. Appl. Ecol.* **2**, 221–239 (1965).

262. R. O. Slatyer and J. V. Lake, Resistance to water transport in plants—whose misconception? *Nature* **212**, 1585–1586 (1966).

263. I. R. Cowan and F. L. Milthorpe, Resistance to water transport in plants—a misconception misconceived. *Nature* **213**, 740–741 (1967).

264. G. V. Parmalee, Applications of meteorological data to indoor climate in buildings. *Bull. Am. Meteorol. Soc.* **36**, 256–264 (1955).

265. P. E. Waggoner and I. Zelitch, Transpiration and the stomata of leaves. *Science* **150**, 1413–1420 (1965).

266. W. M. Rohsenow and H. Choi, "Heat, Mass and Momentum Transfer." Prentice-Hall, Englewood Cliffs, New Jersey, 1961.

267. A. S. Thom, The exchange of momentum, mass, and heat between an artificial leaf and the airflow in a wind-tunnel. *Quart. J. Roy. Meteorol. Soc.* **94**, 44–55 (1968).

268. P. J. Barry and R. E. Munn, Use of radioactive tracers in studying mass transfer in the atmospheric boundary layer. *Phys. Fluids Suppl.* 263–266 (1967).

269. P. M. Bryant, Derivation of working limits for continuous release rates of iodine-131 to atmosphere in a milk producing area. *Health Phys.* **10**, 249–257 (1964).

270. F. Sargent, A dangerous game: taming the weather. *Bull. Am. Meteorol. Soc.* **48**, 452–458 (1967).

271. K. B. Woo, L. Boersma, and L. N. Stone, Dynamic simulation model of the transpiration process. *Water Resources Res.* **2**, 85–97 (1966).

272. W. Ott, J. F. Clarke, and G. Ozolins, Calculating future carbon monoxide emissions and concentrations from urban traffic data. U. S. Dept. of Health, Education, and Welfare, Natl. Center for Air Pollution Control, Cincinnati, Ohio (1967).

273. D. Garfinkel, A simulation study of the effect on simple ecological systems of making rate of increase of population density-dependent. *J. Theoret. Biol.* **14**, 46–58 (1967).

274. L. P. Herrington, On temperature and heat flow in tree stems. Bulletin **73**, School of Forestry, Yale Univ., New Haven, Conn. (1969).

275. H. W. Georgii, Untersuchen über den Luftaustausch zwischen Wohnräumen und Aussenluft. *Arch. Meteorol. Geophys. Bioklimatol.* **B5**, 191–214 (1954).

276. G. R. Lundqvist, Climate-hygienic judgments of occupied rooms based on high accuracy measurements of the carbon dioxide concentration in the stay zone of the room. *In* "Theories of Odors and Odor Measurement" (N. Tanyolac, ed.), pp. 431–445. Robert College Res. Center, Bebek, Istanbul, Turkey, 1968.

277. J. K. S. Wong, Damp cold and damp heat. *Quart. Bull. Div. Mech. Eng. Natl. Aeron. Est.,* Ottawa, pp. 87–98. Natl. Res. Council, Ottawa (1966).

278. A. H. Woodcock, Moisture transfer in textile systems. *Textile Res. J.* **32**, 628–633 (1962).

279. A. C. Burton and O. G. Edholm, "Man in a Cold Environment." Arnold, London, 1955.

280. P. A. Siple and C. F. Passel, Measurements of dry atmospheric cooling in subfreezing temperatures. *Proc. Am. Phil. Soc.* **89**, 177–199 (1945).

281. K. J. K. Buettner, Physical aspects of human bioclimatology. *In* "Compendium of Meteorology" (T. F. Malone, ed.), pp. 1112–1125. Am. Meteorol. Soc., Boston, Massachusetts, 1951.

282. P. M. O. Massey, Finger numbness and temperature in Antarctica. *J. Appl. Physiol.* **14**, 616–620 (1959).

283. O. Wilson, Objective evaluation of wind chill index by records of frostbite in the Antarctica. *Intern. J. Biometeorol.* **11**, 29–32 (1967).

284. E. Flach and W. Mörikofer, Comprehensive climatology of cooling power as measured with the Davos frigorimeter. Part III. Contract DA-91-591-EUC-2848, Meteorol. Observatory, Davos-Platz, Switzerland (1965).

285. O. M. Lidwell and D. P. Wyon, A rapid response radiometer for the estimation of mean radiant temperature in environmental studies. *J. Sci. Inst. Ser. 2* **1**, 534–538 (1968).

286. D. Minard, Prevention of heat casualties in Marine Corps recruits. *Military Med.* **126**, 261–272 (1961).

287. L. G. C. Pugh, Clothing insulation and accidental hypothermia in youth. *Nature* **209**, 1281–1286 (1966).

288. W. H. Portig, The humid warm tropical climate and man. *Weather* **23**, 177–178 (1968).

289. ASHRAE, Handbook of Fundamentals. Am. Soc. Heating, Refrigeration, Air-Conditioning Engrs., New York, 1967.

290. W. Marinov, Die Rolle der Klimatischen Verhältnisse au der Erde bei der Gestaltung des menschlichen Organismus als eines biologischen Systems mit einer Körpertemperatur von 37°C. *Angew. Meteorol.* **5**, 181–183 (1966).

291. J. Chatonnet and M. Cabanac, The perception of thermal comfort. *Intern. J. Biometeorol.* **9**, 182–193 (1965).

292. S. W. Tromp, A physiological method for determining the degree of meteorological cooling. *Nature* **210**, 486–487 (1966).

293. D. H. K. Lee, Physiological instrumentation. Seminar on Human Biometeorology, pp. 25–42. Public Health Service, U. S. Dept. of Health, Education, and Welfare 999-AP-25 (1967).

294. R. E. Huschke, ed., Glossary of Meteorology. Am. Meteorol. Soc., Boston, Massachusetts (1959).

295. E. S. Thom, The discomfort index. *Weatherwise* **12**, 57–60 (1959).

296. R. L. Hendrick, An outdoor weather-comfort index for the summer season in Hartford, Connecticut. *Bull. Am. Meteorol. Soc.* **10**, 620–623 (1959).

297. W. van Beaumont and R. W. Bullard, Sweating: direct influence of skin temperature. *Science* **147**, 1465–1467 (1965).

298. D. H. K. Lee and A. Henschel, Evaluation of thermal environment in shelters. U. S. Dept. of Health Education, and Welfare, Div. Occupational Health, TR-8 (1963).

299. D. H. K. Lee and A. Henschel, Effects of physiological and clinical factors on response to heat. *Ann. N. Y. Acad. Sci.* **134**, 743–749 (1966).

300. H. S. Belding and T. F. Hatch, Index for evaluating heat stress in terms of resulting physiological strains. *Heat Piping Air Conditioning* **27**, 129–136 (1955).

301. P. F. Maycock and B. Matthews, An arctic forest in the tundra of northern Ungava, Quebec. *Arctic* **19**, 114–144 (1966).

302. W. H. Hogg (ed.), Meteorological factors affecting the epidemiology of wheat rusts. Tech. Note No. 99. World Meteorol. Organ., Geneva (1969).

303. P. E. Waggoner, Weather and the rise and fall of fungi. *In* "Biometeorology" (W. P. Lowry, ed.), pp. 45–66. Oregon State Univ. Press, Corvallis, Oregon, 1968.

304. H. Farazdaghi and P. M. Harris, Plant competition and crop yield. *Nature* **217**, 289–290 (1968).

305. L. P. Clark, P. W. Geier, R. D. Hughes, and R. F. Morris, "The Ecology of Insect Populations in Theory and Practice." Methuen, London, 1967.

306. K. E. F. Watt, A computer approach to analysis of data on weather, population fluctuations and disease. *In* "Biometeorology" (W. P. Lowry, ed.), pp. 145–159. Oregon State Univ. Press, Corvallis, Oregon, 1968.

307. E. G. Leigh, On the relation between the productivity, biomass, diversity, and stability of a community. *Proc. U. S. Natl. Acad. Sci.* **53**, 777–783 (1965).

308. S. J. McNaughton, Relationships among functional properties of California grassland. *Nature* **216**, 168–169 (1967).

309. D. R. Margalef, *Mem. Real Acad. Ciencias Artes (Barcelona)* **32**, 373 (1957).

310. D. R. Margalef, Information theory in ecology. *Yearbook Soc. General Systems Res.* **3**, 36–71 (1958).

311. R. P. McIntosh, An index of diversity and the relation of certain concepts to diversity. *Ecology* **48**, 392–404 (1967).

312. M. J. Schroeder, The Hudson Bay high and the spring fire season in the lake states. *Bull. Am. Meteorol. Soc.* **31**, 111–118 (1950).

313. H. Arakawa and J. Tawara, Frequency of air-mass types in Japan. *Bull. Am. Meteorol. Soc.* **30**, 104–105 (1949).

314. H. E. Landsberg, "Physical Climatology" (2nd ed.). Gray Printing, DuBois, Pennsylvania, 1958.

315. W. S. Creswick, Experiments in objective frontal contour analysis. *J. Appl. Meteorol.* **6**, 774–781 (1967).

316. H. H. Lamb, Types and spells of weather around the year in the British Isles: annual trends, seasonal structure of the year, singularities. *Quart. J. Roy. Meteorol. Soc.* **76**, 393–429 (1950).

317. F. Wilmers, Wettertypen für microklimatische Untersuchungen. *Arch. Meteorol. Geophy. Bioklimatol.* **B16**, 144–150 (1968).

318. J. Korshover, Climatology of stagnating anticyclones east of the Rocky Mountains, 1936–1965. Public Health Service Publ. 999-AP-34, Cincinnati, Ohio (1967).

319. G. C. Holzworth, Mixing depths, wind speeds and air pollution potential for selected locations in the United States. *J. Appl. Meteorol.* **6**, 1039–1044 (1967).

320. G. C. Holzworth, Large-scale weather influences on community air pollution potential in the United States. *J. Air Pollution Control Assoc.* **19**, 248–254 (1969).

321. J. D. Stackpole, The air pollution potential forecast program. Weather Bur. Tech. Mem. NMC-43. Natl. Meteorol. Center, Suitland, Maryland (1967).

322. M. E. Miller, Forecasting afternoon mixing depths and transport wind speeds. *Monthly Weather Rev.* **95**, 35–44 (1967).

323. A. C. Stern (ed.), *Proc. Symp. Multiple Source Urban Diffusion Models*, Univ. N. Carolina, Chapel Hill, N. C. (1970) (in press).

324. P. Williams, Air pollution potential over the Salt Lake Valley of Utah as related to stability and wind speed. *J. Appl., Meteorol.* **3**, 92–97 (1964).

325. C. A. Velds, Relation between SO_2 concentration and circulation type in Rotterdam and surroundings. *Preprint Symp. Urban Climates Building Climatol.* World Meteorol. Organ., Geneva (1968).

326. P. Hess and H. Brezowsky, Katalog der Grosswetterlagen Europas. *Ber. Deut. Wetterdienstes* Nr. **33**. Bad Kissingen, Germany (1952).

327. T. Johnson, Rust research in Canada. Publ. 1098, Canada Dept. of Agriculture, Ottawa (1961).

328. J. J. Post (ed.), The influence of weather conditions on the occurrence of apple scab. Tech. Note No. 55. World Meteorol. Organ., Geneva (1963).

329. W. H. Hogg, The use of upper air data in relation to plant disease. *In* "Weather and Agriculture" (J. A. Taylor, ed.), pp. 115–127. Pergamon Press, Oxford, 1967.

330. K. R. Petersen, Continuous point source plume behavior out to 160 miles. *J. Appl. Meteorol.* **7**, 217–226 (1968).

331. M. S. Hirt and S. E. Dinning, Experiment in pollution transport during Peel Country Cleaner Air Week campaign. *Atmosphere* **7**, 70–72 (1969).

332. K. J. Richardson, Discussion, *Quart. J. Roy. Meteorol. Soc.* **85**, 171–176 (1959).

333. M. W. Shaw, The diamond-back moth migration of 1958. *Weather* **17**, 221–234 (1962).

334. G. W. Hurst, The 1962 invasion of the small willow moth in Southern England. Unpublished manuscript. Meteorol. Office, Bracknell, England (1963).

335. R. C. Rainey, Discussion. *Quart. J. Roy Meteorol. Soc.* **85**, 171–176 (1959).

336. C. G. Johnson, The vertical distribution of aphids in the air and the temperature lapse rate. *Quart. J. Roy. Meteorol. Soc.* **83**, 194–201 (1957).

337. R. C. Rainey, Meteorology and the migration of desert locusts. Tech. Note No. 54. World Meteorol. Organ., Geneva (1963).

338. J. Roffey and G. Popov, Environmental and behavioural processes in a desert locust outbreak. *Nature* **219**, 446–449 (1968).

339. J. S. Kennedy, A turning point in the study of insect migration. *Nature* **189**, 785–791 (1961).

340. R. C. Rainey, Some observations of flying locusts and atmospheric turbulence in eastern Africa. *Quart. J. Roy. Meteorol. Soc.* **84**, 334–354 (1958).

341. D. E. Pedgley and P. M. Symmons, Weather and the locust upsurge. *Weather* **23**, 484–492 (1968).

342. R. S. Scorer, Notes on middle-scale motions affecting locust swarms. Meteorology and the Desert Locust, pp. 265–273. Tech. Note No. 69, World Meteorol. Organ., Geneva (1965).

343. R. C. Rainey, Notes on current trends in locust meteorology and related topics. C. Ag. M.-IV, Doc. 13, Appendix B, World Meteorol. Organ., Geneva (1967).

344. F. A. Huff, Relation between leafhopper influences and synoptic weather conditions. *J. Appl. Meteorol.* **2**, 39–43 (1963).

345. A. M. Bagg, Factors affecting the occurrence of the Eurasian lapwing in Eastern North America. *Living Bird* **6**, 87–122 (Cornell Lab. Ornithology) (1967).

346. R. F. Reinking, Insolation reduction by contrails. *Weather* **23**, 171–173 (1968).

347. M. G. Pearson, Insolation reduction by contrails. *Weather* **23**, 520 (1968).

348. R. F. Reinking, Insolation reduction by contrails. *Weather* **23**, 520–521 (1968).

349. M. L. Nicodemus and J. D. McQuigg, A simulation model for studying possible modification of surface temperature. *J. Appl. Meteorol.* **8**, 199–204 (1969).

350. Weather forecasting of aircraft condensation trails. Air Weather Service Manual, pp. 100–105. U. S. Air Force, Scott Air Force Base, Illinois (1960).

351. J. P. Bruce and R. H. Clark, "Introduction to Hydrometeorology." Pergamon Press, Oxford (1966).

352. R. A. Pullen, Computation of the moisture content of the atmosphere using surface level climatological data. *J. Hydrol.* **6**, 168–182 (1968).

353. F. Singleton, Estimates of extreme precipitation. *Meteorol. Mag.* **96**, 350–351 (1967).
354. V. Yevjevich, Misconceptions in hydrology and their consequences. *Water Resources Res.* **4**, 225–232 (1968).
355. J. G. Lockwood, Extreme rainfalls. *Weather* **23**, 284–289 (1968).
356. B. Lettau, The transport of moisture into the antarctic interior. *Tellus* **21**, 331–340 (1969).
357. D. M. Driscoll and H. E. Landsberg, Synoptic aspects of mortality. A case study. *Intern. J. Biometeorol.* **11**, 323–328 (1967).
358. F. Sargent, Changes in ideas on the climatic origin of disease. *Bull. Am. Meteorol. Soc.* **41**, 238–244 (1960).
359. S. J. Carne, Study of the effect of air pollution upon respiratory diseases in London in the winters of 1962–63 and 1963–64. *Proc. Intern. Cong. Clean Air,* London pp. 259–261. Natl. Soc. for Clean Air (1966).
360. B. J. Garnier, Weather conditions in Nigeria. Climatol. Res. Series No. 2, Dept. of Geography, McGill Univ., Montreal, Quebec (1967).
361. B. R. Bean and Q. L. Florey, A field study of the effectiveness of fatty alcohol mixtures as evaporation reducing monomolecular films. *Water Resources Res.* **4**, 206–208 (1968).
362. L. J. Fritschen and P. R. Nixon, Microclimate before and after irrigation. Ground Level Climatology, pp. 351–366, Publ. No. 86. Am. Assoc. Advan. Sci., Washington, D. C. (1965).
363. E. Vowinckel, Evaporation on the Canadian prairies. Publ. in Meteorol. No. 88. Dept. of Meteorol., McGill Univ., Montreal, Quebec (1967).
364. M. Gangopadhyaya (ed.), Measurement and estimation of evaporation and evapotranspiration. Tech. Note No. 83. World Meteorol. Organ., Geneva (1966).
365. S. L. Yu and W. Brutsaert, Evaporation from very shallow pans. *J. Appl. Meteorol.* **6**, 265–271 (1967).
366. G. E. Harbeck, Jr., A practical technique for measuring reservoir evaporation utilizing mass transfer theory. U. S. Geol. Survey, Prof. Paper 272-E, pp. 101–105 (1962).
367. C. W. Thornthwaite, The climates of North America according to a new classification. *Geograph. Rev.* **21**, 633–655 (1931).
368. C. W. Thornthwaite, An approach toward a rational classification of climate. *Geograph. Rev.* **38**, 55–94 (1948).
369. W. L. Pelton, K. M. King, and C. B. Tanner, An evaluation of the Thornthwaite and mean temperature methods for determining potential evapotranspiration. *Agron. J.* **52**, 387–395 (1960).
370. U. Högström, Studies of the natural evaporation and energy balance. *Tellus* **20**, 65–75 (1968).
371. H. L. Penman, Vegetation and hydrology, Tech. Commun. No. 53. Commonwealth Bureau of Soils, Harpenden, England (1963).
372. J. W. Holmes and C. L. Watson, The water budget of irrigated pasture land near Murray Bridge, South Australia. *Agr. Meteorol.* **4**, 177–188 (1967).
373. C. B. Tanner and M. Fuchs, Evaporation from unsaturated surfaces: a generalized combination method. *J. Geophys. Res.* **73**, 1299–1304 (1968).
374. W. O. Prouitt and F. J. Lourence, Correlation of climatological data with water requirements of crops. Water Sci. Eng. Papers 9001. Univ. of California, Davis, California (1968).

375. L. A. Hunt, I. I. Impens, and E. R. Lemon, Estimates of the diffusion resistance of some large sunflower leaves in the field. *Plant Physiol.* **43**, 522–526 (1968).

376. G. Szeicz and I. F. Long. Surface resistance of crop canopies. *Water Resources Res.* **5**, 622–633 (1969).

377. C. H. B. Priestley, "Turbulent Transfer in the Lower Atmosphere." Univ. of Chicago Press, Chicago, Illinois (1959).

378. R. E. Munn and E. J. Truhlar, Potential evapotranspiration (manuscript in preparation) (1970).

379. S. F. Fedorov, Evaporation from forest and field in years with different amounts of moisture. *Trans. State Hydrol. Inst. (Trudy GGI)* **123**, 22–35; *Trans. State Hydrolog. Inst.* (English Transl. by Am. Geophys. Union in *Soviet Hydrol.* **4**, 337–348) (1965).

380. W. Baier and G. W. Robertson, A new versatile soil moisture budget. *Can. J. Plant Sci.* **46**, 299–315 (1966).

381. W. Baier, Concepts of soil moisture availability and their effect on soil moisture estimates from a meteorological budget. *Agr. Meteorol.* **6**, 165–178 (1969).

382. G. W. Smith, The relation between soil moisture and incidence of disease in cacao. *J. Appl. Meteorol.* **2**, 614–618 (1963).

383. R. F. Dale and R. H. Shaw, The effect on corn yields of moisture stress and stand at two fertility levels. *Agron. J.* **57**, 475–479 (1965).

384. R. F. Dale and R. H. Shaw, The climatology of soil moisture, atmospheric evaporative demand, and resulting moisture stress days for corn at Ames, Iowa. *J. Appl. Meteorol.* **4**, 661–669 (1965).

385. G. A. McKay, J. Maybank, O. R. Mooney, and W. L. Pelton, The agricultural climate of Saskatchewan. Climatol. Studies No. 10. Meteorol. Branch, Toronto (1967).

386. W. C. Palmer, The abnormally dry weather of 1961–66 in the Northeastern United States. *Proc. Conf. Drought Northeastern United States* pp. 32–56. New York Univ., Geophys. Res. Lab. TR-68-3 (1968).

387. V. P. Subrahmanyam, Incidence and spread of continental drought. WMO/IHD Rept. No. 2. World Meteorol. Organ., Geneva (1967).

388. W. C. Palmer, Meteorological drought. Res. Paper No. 45. U. S. Weather Bureau (1964).

389. D. J. Fieldhouse and W. C. Palmer, Meteorological and agricultural drought. Bull. No. 353. Univ. of Delaware, Agr. Exp. Station, Newark, Delaware (1965).

390. W. C. Palmer, Keeping track of crop moisture conditions, nationwide: the new crop moisture index. *Weatherwise* **21**, 156–161 (1968).

391. J. Namias, Nature and possible causes of the Northeastern United States drought during 1962–65. *Monthly Weather Rev.* **94**, 543–554 (1966).

392. A. R. Sen, A. K. Biswas, and D. K. Sanyal, The influence of climatic factors on the yield of tea in the Assam valley. *J. Appl. Meteorol.* **5**, 789–800 (1966).

393. S. A. Changnon and J. C. Neill, A mesoscale study of corn-weather response on cash-grain farms. *J. Appl. Meteorol.* **7**, 94–104 (1968).

394. F. A. Huff and S. A. Changnon, Evaluation of cloud seeding benefits to corn yields. *Preprint Natl. Conf. Agr. Meteorol., 8th* Am. Meteorol. Soc., Boston (1968).

395. M. S. Kulik, The quantity and quality of crop yields. WMO Commission Agr. Meteorol., 4th Session, Doc. 16, World Meteorol. Organ., Geneva (1967).

396. W. P. Lowry, Biometeorological inference and plant-environment interrelations.

Ground Level Climatology, pp. 3–14, Publ. No. 86. Am. Assoc. Advan. Sci., Washington, D. C. (1967).

397. H. Arakawa, Three great famines in Japan. *Weather* **12**, 211–217 (1957).

398. W. L. Pelton, The effect of a windbreak on wind travel, evaporation and wheat yield. *Can. J. Plant Sci.* **47**, 209–214 (1967).

399. W. Baier and G. W. Robertson, Estimating yield components of wheat from calculated soil moisture. *Can. J. Plant Sci.* **47**, 617–630 (1967).

400. W. Baier and G. W. Robertson, The performance of soil moisture estimates as compared with the direct use of climatological data for estimating crop yields. *Agr. Meteorol.* **5**, 17–31 (1968).

401. L. P. Smith, Forecasting annual milk yields. *Agr. Meteorol.* **5**, 209–214 (1968).

402. G. W. Hurst, Honey production and summer temperatures. *Meteorol. Mag.* **96**, 116–120 (1967).

403. B. Tom, M. Brown, and R. Chang, Peptic ulcer disease and temperature changes in Hawaii. *J. Appl. Meteorol.* **3**, 311–315 (1964).

404. S. Licht (ed.), "Medical Climatology." Waverly Press, Baltimore, Maryland, 1964.

405. F. H. Schmidt and C. A. Velds, On the relation between changing meteorological circumstances and the decrease of concentration of sulphur dioxide around Rotterdam. *Atmos. Environ.* **3**, 455–460 (1969).

406. P. A. Sheppard, Preface. "World Climate from 8000 to 0 BC" (J. S. Sawyer, ed.), p. 1. Roy. Meteorol. Soc., London, 1966.

407. G. Manley, The mean temperature of central England, 1698–1952. *Quart. J. Roy. Meteorol. Soc.* **79**, 242–261 (1953).

408. H. E. Landsberg, The decennial United States census of climate 1960 and its antecedents; key to meteorological records documentation. No. 6.2, U. S. Weather Bureau, Washington, D. C. (1960).

409. H. H. Lamb, What can we find out about the trend of our climate? *Weather* **18**, 194–216 (1963).

410. J. Oliver, A weather record of the early eighteenth century. *Weather* **16**, 335–340 (1961).

411. R. A. Bryson and P. R. Julian (eds.), *Proc. Conf. Climate Eleventh Sixteenth Centuries* NCAR Tech. Note 63-1, Natl. Center Atmospheric Res., Boulder, Colorado (1963).

412. D. Thomas, Climate and cropping in the early nineteenth century in Wales. *In* "Weather and Agriculture" (J. A. Taylor, ed.), pp. 201–212. Pergamon Press, Oxford, 1967.

413. J. Oliver, Problems of agro-climatic relationships in Wales in the eighteenth century. *In* "Weather and Agriculture" (J. A. Taylor, ed.), pp. 187–200. Pergamon Press, Oxford, 1967.

414. J. Oliver, Discussion. *In* "The Biological Significance of Climatic Changes in Britain" (C. G. Johnson and L. P. Smith, eds.). Academic Press, New York, 1965.

415. G. Manley, Climatic variation. *Quart. J. Roy. Meteorol. Soc.* **79**, 185–209 (1953).

416. H. Godwin, Introductory address. "World Climate from 8000 to 0 BC" (J. S. Sawyer, ed.), pp. 1–14. Roy. Meteorol. Soc., London, 1966.

417. H. E. Suess, Climatic changes, solar activity, and the cosmic-ray production rate of natural radiocarbon. *Meteorol. Monogr.* **8**(30), 146–150 (1968).

418. H. H. Lamb, In discussion of trees and climatic history of Scotland. *Quart. J. Roy. Meteorol. Soc.* **91**, 546–547 (1965).

419. V. Mitchell, An investigation of certain aspects of tree growth rates in relation to climate in the central Canadian boreal forest. Tech. Rept. No. 33. Dept. of Meteorol., Univ. of Wisconsin (1967).

420. H. C. Fritts, Growth-rings of trees: their correlation with climate. *Science* **154**, 973–979 (1966).

421. B. Seddon, Prehistoric climate and agriculture: a review of recent paleo-ecological investigations. In "Weather and Agriculture" (J. A. Taylor, ed.), pp. 173–185. Pergamon Press, Oxford, 1967.

422. G. Manley, Problems of the climatic optimum: the contribution of glaciology. In "World Climate from 8000 to 0 BC" (J. S. Sawyer, ed.), pp. 34–39. Roy. Meteorol. Soc., London, 1966.

423. J. T. Bailey, S. Evans, and G. de Q. Robin, Radio echo sounding of polar ice sheets. *Nature* **204**, 420–421 (1964).

424. S. Evans, Discussion. *Quart. J. Roy. Meteorol. Soc.* **91**, 545 (1965).

425. J. D. H. Wiseman, Evidence for recent climatic changes in cores from the ocean bed. In "World Climate from 8000 to 0 BC" (J. S. Sawyer, ed.), pp. 84–97. Roy. Meteorol. Soc., London, 1966.

426. C. H. Hendy and A. T. Wilson, Palaeoclimatic data from speleotherms. *Nature* **219**, 48–51 (1968).

427. J. Emiliani, Pleistocene temperatures. *J. Geol.* **63**, 538–578 (1955).

428. S. K. Runcorn, Geophysics and palaeoclimatology. *Quart. J. Roy. Meteorol. Soc.* **91**, 257–267 (1965).

429. L. Starkel, Post-glacial climate and the moulding of European relief. In "World Climate from 8000 to 0 BC" (J. S. Sawyer, ed.), pp. 15–32. Roy. Meteorol. Soc., London, 1966.

430. H. H. Lamb, Britain's changing climate. In "The Biological Significance of Climatic Changes in Britain" (C. G. Johnson and L. P. Smith, eds.), pp. 3–31. Academic Press, New York, 1965.

431. H. C. Hoinkes, Glacier variation and weather. *J. Glaciol.* **7**, 3–19 (1968).

432. E. Wahl, C. Behrens, and B. Hayden, Project Frontier Fort. Annual Summary, 1966–67, Interdisciplinary program of climatic research (R. A. Bryson, ed.), pp. 5–6. Dept. of Meteorol., Univ. of Wisconsin, Madison, Wisconsin (1967).

433. E. Lorenz, Climatic determinism. *Meteorol. Monogr.* **8**, No. 30, 1–3 (1968).

434. J. A. Larsen, The vegetation of the Ennadai Lake area NWT: studies in sub-arctic and arctic climatology. *Ecolog. Monogr.* **35**, 37–59 (1965).

435. K. Takahashi, Climatic changes calculated by a simple heat transfer model at the earth surface. *J. Meteorol. Soc. Japan* **43**, 188–195 (1965).

436. G. R. Lewthwaite, Environmentalism and determinism: a search for clarification. *Ann. Assoc. Am. Geograph.* **56**, 1–23 (1966).

437. R. A. Bryson, The research frontier. *Saturday Rev.* **Apr. 1**, 52–55 (1967).

438. R. A. Bryson and D. A. Baerreis, Possibilities of major climatic modification and their implications: northwest India, a case for study. *Bull. Am. Meteorol. Soc.* **48**, 136–142 (1967).

439. S. A. Barnett, Mice at −3°C. *New Scientist* **27**, 678–679 (1965).

440. A. Court, Climatic classification and plant geography in 1842. *Weather* **22**, 276–288 (1967).

441. W. Köppen, "Grundriss der Klimakunde," 2nd ed. W. de Gruyter, Berlin and Leipzig, 1931.

442. E. Suzuki, A statistical and climatological study on the rainfall in Japan. *Papers Meteorol. Geophys.* **18**, 103–181 (1967).

443. B. Weisman, D. H. Matheson, and M. Hirt, Air pollution survey for Hamilton, Ontario. *Atmos. Environ.* **3,** 11–23 (1969).

444. R. O. Weedfall and B. Linsky, A mesoclimatological classification system for air pollution engineers. *J. Air Pollution Control Assoc.* **19,** 511–513 (1969).

445. W. P. Lowry and H. E. Reiquam, An index for analysis of the buildup of air pollution potential. *Preprint Annual Meeting. Air Pollution Control Assoc., 61st, Pittsburgh, Pennsylvania* (1968).

446. I. A. Singer and M. E. Smith, Relation of gustiness to other meteorological parameters. *J. Meteorol.* **10,** 121–126 (1953).

447. P. M. Bryant, Methods of estimation of the dispersion of windborne material and data to assist in their application. U. K. At. Energy Authority, Harwell, England (1964).

448. W. D. Sellers, "Physical Climatology." Univ. of Chicago Press, Chicago, Illinois, 1965.

449. M. I. Budyko, The Heat Balance of the Earth's Surface (English transl. from the Russian). U. S. Dept. of Commerce, Washington, D. C., 1958 (1956).

450. H. H. Lettau, Evapotranspiration Climatonomy. *Monthly Weather Rev.* **97,** 691–699 (1969).

451. C. W. Thornthwaite and J. R. Mather, Average climatic water balance data of the continents. Pt. I Africa. *Publications Climatol.* **15,** No. 2. Drexel Inst. of Technol., Centerton, New Jersey (1962).

452. V. P. Subrahmanyam and C. V. S. Sastri, Some aspects of drought climatology of the dry subhumid zones of south India. *J. Meteorol. Soc. Japan* **47,** 239–244 (1969).

453. D. L. Gary, Project Creosote. Annual Summary, 1966–67, Interdisciplinary program of climatic research (R. A. Bryson, ed.), pp. 61–62. Dept. Meteorol., Univ. Wisconsin, Madison, Wisconsin (1967).

454. G. P. deBrichambaut and C. C. Wallén, A study of agroclimatology in semi-arid and arid zones of the Near East. WMO Tech. Note No. 56 (1963).

455. R. J. Kopec, Continentality around the Great Lakes. *Bull. Am. Meteorol. Soc.* **46,** 54–57 (1965).

456. V. M. Polowchak and H. A. Panofsky, The spectrum of daily temperatures as a climatic indicator. *Monthly Weather Rev.* **96,** 596–600 (1968).

457. Guide to Agricultural Meteorological Practices. No. 134. TP 61. World Meteorol. Organ., Geneva (1963).

458. J. J. Burgos, Agroclimatic classifications and representations. WMO Commission Agr. Meteorol., 2nd Session, Document 18, Warsaw (1958).

459. I. Hustich, On variations in climate, in crop of cereals and in growth of time in northern Finland 1890–1939. *Fennia* **70,** 2, 1–24 (1947).

460. W. H. Terjung, Annual physioclimatic stresses and regimes in the United States. *Geograph. Rev.* **57,** 225–240 (1967).

461. H. P. Bailey, Towards a unified concept of the temperate climate. *Geograph. Rev.* **54,** 516–545 (1964).

462. C. E. Hounam, Climate and air conditioning requirements in sparsely occupied areas of Australia. *Preprint Symp. Urban Climates Building Climatol.* World Meteorol. Organ., Geneva (1968).

463. H. V. Foord, An index of comfort for London. *Meteorol. Mag.* **97,** 282–286 (1968).

464. M. Gregorczuk, Bioclimates of the world related to air enthalpy. *Intern. J. Biometeorol.* **12,** 35–39 (1968).

465. W. J. Maunder, A human classification of climate. *Weather* **17**, 3–12 (1962).
466. N. E. Davis, An optimum summer weather index. *Weather* **23**, 305–317 (1968).
467. F. Prohaska, Climatic classifications and their terminology. *Intern. J. Biometeorol.* **11**, 1–3 (1967).
468. W. R. D. Sewell, R. W. Kates, and L. E. Phillips, Human response to weather and climate. *Geograph. Rev.* **58**, 262–280 (1968).
469. D. A. Mazzarella, Sometimes on Sunday—it rains. *Weatherwise* **20**, 259–263 (1967).
470. J. K. Page, The fundamental problems of building climatology considered from the point of view of decision making by the architect and urban designer. *Preprint Symp. Urban Climates Building Climatol., Brussels, CLU/Document 19* World Meteorol. Organ., Geneva (1968).
471. I. I. Gringorten, Forecasting by statistical inferences. *J. Meteorol.* **7**, 388–394 (1950).
472. B. Shorr, The cost/loss utility ratio. *J. Appl. Meteorol.* **5**, 801–803 (1966).
473. J. C. Thompson, Weather decision making—the pay-off. *Bull. Am. Meteorol. Soc.* **44**, 75–78 (1963).
474. L. E. Borgman, Weather-forecast profitability from a client's viewpoint. *Bull. Am. Meteorol. Soc.* **41**, 374–356 (1960).
475. E. S. Epstein, A Bayesian approach to decision making in applied meteorology. *J. Appl. Meteorol.* **1**, 169–177 (1962).
476. H. Arakawa, Usefulness of weather forecasting and storm warnings. *Weather* **21**, 46–47 (1966).
477. B. J. Mason, The role of meteorology in the national economy. *Weather* **21**, 382–393 (1966).
478. L. S. Gaudin and R. L. Kagan, The economic approach to the planning of a network of meteorological stations. *Trans. Voeikov Main Geopyhs. Observatory Tr. GGO* **208**, 120–131 (English transl. *Soviet Hydrol.* **6**, 597–606, Am. Geophys. Union) (1967).
479. R. S. Edwards, Economic measurement of weather hazards. *Weather* **24**, 70–73 (1969).
480. H. G. de Wiljes and J. C. A. Zaat, The influence of climate upon the number of weather-working hours in combine harvesting in the Netherlands. *Arch. Meteorol. Geophys. Bioklimatol.* **B16**, 105–114 (1968).
481. L. P. Smith, Meteorology applied to agriculture. *WMO Bull.* **16**, 190–194 (1967).

INDEX